An Introduction to PHOTOBIOLOGY

An Introduction to
PHOTOBIOLOGY

A series of introductory lectures
given under the auspices of the
Fifth International Congress on Photobiology
at Dartmouth College, Hanover, New Hampshire,
August 26–31, 1968

CARL P. SWANSON, *Editor*

Prentice-Hall, Inc.
Englewood Cliffs, New Jersey

PRENTICE-HALL INTERNATIONAL, INC., *London*
PRENTICE-HALL OF AUSTRALIA, PTY. LTD., *Sydney*
PRENTICE-HALL OF CANADA, LTD., *Toronto*
PRENTICE-HALL OF INDIA PRIVATE LIMITED, *New Delhi*
PRENTICE-HALL OF JAPAN, INC., *Tokyo*

Current printing (last digit):
10 9 8 7 6 5 4 3 2 1

13-492421-5

Library of Congress Catalog Card Number: 72-83357

Printed in the United States of America

Foreword

Photobiology differs from most of the subdisciplines of biology in that its limitations are set not by biological, but by rather narrow physical, parameters. This would seem, at first glance, to provide circumstances for the development of an esoteric science appealing to a small number of specialists, with their commonality of interest centered principally upon a limited portion of the electromagnetic spectrum. However, the reverse situation seems to be true and is attributable to a number of factors. The pervasive influence of visible and ultraviolet light on biological systems of all degrees of simplicity, complexity, and phylogenetic position; the fact that initial absorption of light by the chromophoric groups of crucial molecules leads to responses detectable at levels ranging from the singlet and triplet states of simple molecules to the behavior of intact organisms; the observation that reaction times span a scale from fractions of seconds to the seasonal migration of birds: all these are such as to bring together for common discussion, under the aegis of photobiology, a highly diversified group of scientists. Their diversity finds expression in training, techniques, outlook, biological material, and measurable reaction responses; their unity relates to the interactions of light and life.

No scientist, by training or inclination, can function experimentally and critically in all of the heterogeneous areas of photobiology. If for no other reason, the simple prerequisite of knowing well one's experimental material would be a deterrent; the scientific equivalent of the Renaissance man does not exist in this day of required specialization. Nevertheless, the photobiologists have much to learn from each other, particularly since the field is so unstructured as to have avoided so far the compartmentalization that interferes with free communication. The quadrennial congress becomes, therefore, not only a means for the presentation of research results and the renewal of acquaintances but also, when properly managed, an effective means of instruction within the group.

With this thought in mind, the Program Subcommittee of the United States Organizing Committee planned a series of introductory lectures to be presented at the Fifth International Congress of Photobiology held at Dartmouth College, Hanover, New Hampshire, on August 26–31, 1968. The lecturers were chosen not only for their eminence within their respective fields, but also for their lecturing ability. They were instructed to make an intelligible and interesting presentation, not to a group of their scientific peers, but to a general audience of photobiologists. Their papers do not constitute a survey, but rather provide an introduction to segments of the broad, varied, dynamic, and enlarging discipline of photobiology. Seven papers had been planned for publication. However, the conclusions by Dr. Colin S. Pittendrigh in his lecture on circadian rhythms have been superseded by subsequent investigations. Consequently, his paper will not be included in this volume. The original intention of the Program Subcommittee was one of self-enlightenment, but it is hoped that other biologists will find the lectures of interest. Possibly a student may be photoreactivated by them, enter the field of photobiology, and become a future contributor.

The beginnings of photobiology are lost in the abyss of history. However, through the courtesy of Dr. Holger Brodthagen of the Finsen Institute, Copenhagen, Denmark, it has been possible to illustrate as a frontispiece one of the first photobiological experiments. It occurred sometime during the 14th century B.C. Pictured are Akhenaten, a pharaoh of the 18th Dynasty; his wife Nefertiti; and their three princesses. Known as the "Manifesto" of the solar cult of Amarna, this illustration in stone shows some of the rays from the sun terminating in the left hand of Akhenaten in which is held the Ankh, symbol of life. It can be safely assumed that Akhenaten recognized in his own way the relation between light and life. Born Amenhoteph IV, he is reputed to have initiated a monotheistic religion with Aton, the sun, as God. In recognition of this, he changed the prefix Amen to the suffix Aton or Aten and renamed himself Akhenaten. One of his daughters became the wife of King Tut Ankamon, the spelling of whose name indicates a return of the old religion, termination of the sun cult, and temporary eclipse of photobiology among the pharaohs. The present group of lectures do not necessarily celebrate a return to the worship of light and life, but they do give an indication of our present level of understanding of the role of light in biological processes.

C. P. SWANSON, *Editor*

Contents

An Introduction to
PHOTOBIOLOGY

Photochemistry of Complex Molecules

George Porter, F.R.S.

Davy Faraday Research Laboratory of the Royal Institution, London

HISTORICAL

As is to be expected for phenomena which are so widespread and so fundamentally important a part of our lives, the study of photobiology can claim to have beginnings as early as those of any science. The writer of the book of Genesis seems to have appreciated that without light there can be no life when he recorded, in the third verse of the Bible, that the primary event of the creation was the appearance of light. The Greeks recognized that light may play a part in biochemical change: Aristotle reported that light is necessary for photosynthesis to occur, and Herodotus went a step too far when he said, "Exposure to the sun is eminently necessary to those who are in need of building themselves up and putting on weight"!

The phenomena of phosphorescence and chemiluminescence have long been known, and, as early as 1667,

Robert Boyle showed that oxygen was necessary for the appearance of bioluminescence. The nineteenth century saw the introduction of photography, the rediscovery by Draper of the law of Grotthus, that only light which is absorbed can cause chemical change, and the classical studies of the hydrogen-chlorine reaction by Bunsen and Roscoe. The quantum theory, discovered in 1900, enabled Einstein to introduce his law of photochemical equivalence eight years later, which in turn led to the all important concept of the quantum yield, pioneered by Warburg.

If it can be said to have had a beginning at all, organic photochemistry and the photochemistry of complex molecules in general started with the systematic studies of Ciamician at the beginning of this century, and a number of the organic photochemical reactions which are still of greatest interest today were first discovered by him. It is remarkable that half a century should pass with so little activity or progress in the field after his work. The interval between the world wars saw important advances in the study of photosynthesis and was a very active period in the field of photochemical gas kinetics. It was also a period of great activity in molecular spectroscopy and the electronic structure of complex molecules which was to provide the basis for the rapid developments in the photochemistry of complex molecules which have taken place over the last decade. These developments have resulted from the convergence of many disciplines. We owe our present understanding of the primary photophysical processes in complex molecules largely to studies of luminsecence and spectroscopy, improved instrumentation in this field, and the introduction of techniques such as direct fluorescence lifetime measurement and flash photolysis. There has also been a greatly increased interest in photochemistry as a preparative tool for the organic chemist and as the basis of photobiology. It is not surprising if such a variety of disciplines does not yet have a common language, and there can be few subjects in which an international conference can be more profitable.

LAWS AND GENERALIZATIONS OF PHOTOCHEMISTRY

Nearly every theoretical and experimental aspect of the photochemistry of complex molecules is to be reviewed, more expertly and in more detail, during this conference. In this introductory paper I shall merely try to summarize some of the generalizations which seem to be emerging and which are valid, with few exceptions, over the whole field of photochemistry and photobiology. I shall resist the temptation to state any of these principles as "Laws of Photochemistry." Laws such as those of Grotthus and Draper and of Einstein are derivative from other branches of science and are, therefore, somewhat unnecessary as separate statements. Furthermore, it is difficult to state the Einstein law in a concise, useful form while taking into account such phenomena as biphotonic processes. Exceptions may even be found, in principle, to the Grotthus and Draper law; for example, a photon may stimulate emission from an excited molecule and so change its chemical behavior without being absorbed. On the other hand, one occasionally meets reports in the literature of photochemical effects in systems where there is no absorption of the pure substances described at the wavelength of the light source used, so the principle of Gotthus and Draper is one of which we need to be reminded from time to time.

A photochemical reaction is one which occurs initially from an electronically excited state of the molecule. If this state is thermally equilibrated; then the principles which determine the course of the chemical reaction are essentially the same as those which determine chemical reactivity in the ground state, though the excited state must be regarded as a new chemical species with its own characteristic electronic distribution and chemical reactivity. Generalizations regarding photochemical behavior are conveniently discussed in two categories: (a) The photophysical processes which precede chemical change and (b) the chemical reactivity as a function of different types of electronic states of a molecule.

In order to exclude a number of phenomena which are not usually of interest in photobiology, the discussion will be confined, except where specifically stated otherwise, to complex molecules in condensed phases. Fortunately, this restriction is a considerable simplification and excludes many complexities which occur with simpler molecules in the gas phase. The size of a molecule which can be regarded as complex in this sense is not precisely known but can be arbitrarily, and probably fairly safely, set at ten or more atoms. A further simplification, which is a far less realistic one for biological systems, is that we shall, for the present, restrict discussion to systems sufficiently dilute to exclude solute-solute interactions. The solute molecule is assumed to have a ground state of singlet multiplicity.

PHOTOPHYSICAL PROCESSES

The physical processes which follow the absorption of a light quantum are discussed in terms of the Jablonski diagram, as extended to singlet and triplet manifolds of states by Lewis and Kasha, and by Terenin (Fig. 1). The lines of this diagram are the states on which we can write a wide variety of photochemical tunes. The notation is well known. The heavy horizontal bars on the left represent the singlet electronic energy levels, S_1, S_2, including the lowest one, S_0, which is the ground state; those on the right represent the triplet levels, T_1, T_2, and the lighter bars represent the vibrationally excited levels associated with each electronically excited state. The relative rates of conversion between these levels determine the course of chemical change and the luminescence properties; continuing our musical analogy, we may say that the thin staves hold quavers and the thick ones crotchets, except for the lowest which hold minims, though the relative time value of these notes is nearer to decimal than to binary ratios. We must consider the rates of these processes a little more precisely, although, since we seek generalizations, our quantities will necessarily be orders of magnitude at the best.

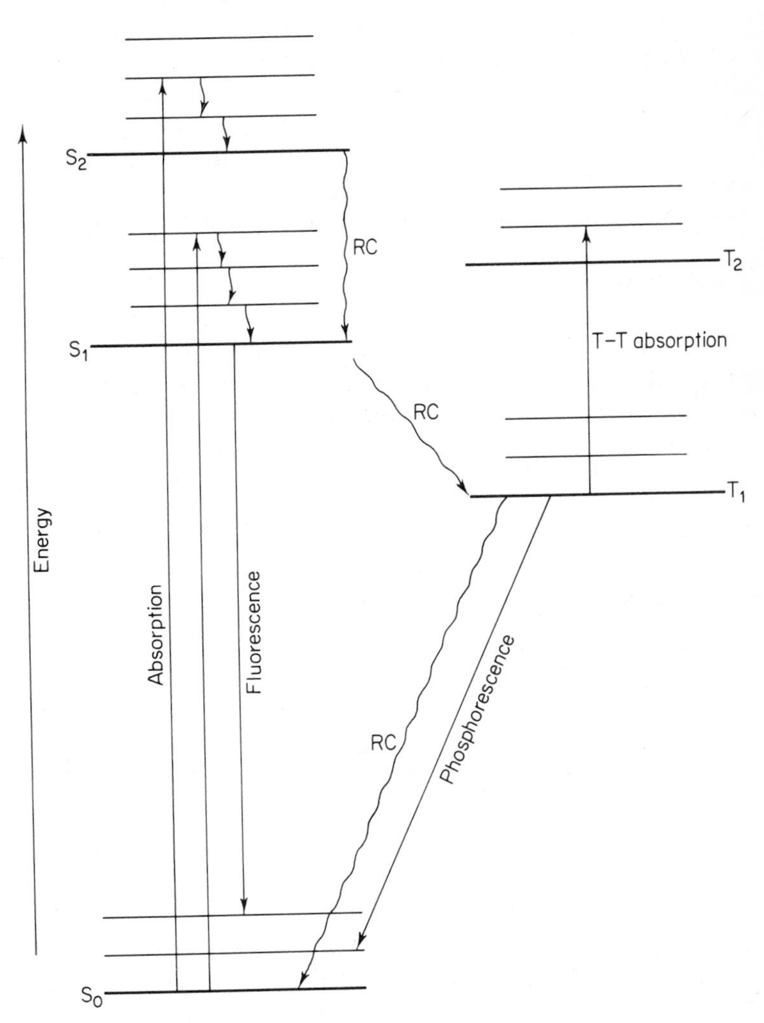

Figure 1

Vibrational Relaxation:
Rate = 10^{11} to 10^{13} Sec^{-1}

The lower limit is set by the absence of observable resonance fluorescence from states which show strong emission from the lowest vibrational levels; the upper limit is set by the uncertainty principle and the fact that vibrational structure is usually observed in the absorption spectrum. Recently, by use of modal locked lasers, Rentzepis has measured the rate of vibrational relaxation directly in the lowest excited singlet state of azulene and found a relaxation time of 8 picosec (1). There is good reason to suppose that all complex molecules in solution will have similar relaxation times.

Radiationless Conversion

These are the processes marked RC in Fig. 1, and they occur by the two-stage mechanism represented in Fig. 2. Owing to the higher density of states in the

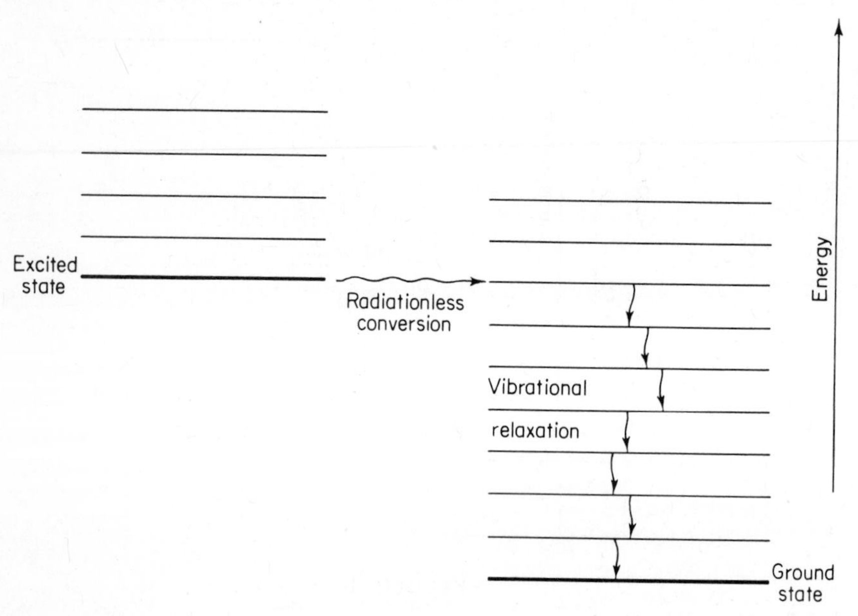

Figure 2

isoenergetic vibrationally excited lower state, the reverse crossing probably never occurs before vibrational relaxation, and the radiationless conversion itself is, therefore, always rate determining. This is the process which has been discussed by Wilse Robinson (2) whose expression for the rate in its simplest form leads to the approximate proportionality for the rate constant k:

$$k \propto \beta_{el}^2\ (\mathrm{cm}^{-1}) \langle \phi''/\phi' \rangle^2\ \mathrm{sec}^{-1}$$

Typical values of β_{el}, the electronic part of the matrix element of perturbation between the two states, are 100 cm^{-1} for allowed transitions and 0.01 cm^{-1} for spin forbidden transitions. The term $\langle \phi''/\phi' \rangle$ is the vibrational part of the matrix element and depends on the overlap of vibrational wave functions of the two states, a Frank-Condon factor which decreases with the energy separation of the two electronic states. As a result of this, radiationless conversion probabilities differ widely, and are not readily calculated, and few generalizations are possible. However, empirical evidence from the study of a wide range of molecules leads to the following useful statements about rates which, although they may be modified in the light of further work, seem to represent the situation for the majority of complex molecules.

Radiationless Conversion Rates

(a) Between two excited states of the same multiplicity. Rate > 10^{10} sec^{-1}. The evidence for this figure is the low, usually unobservable yield of luminescence from all states other than S_1 and T_1. The conversion rate of the T_2 state of anthracene derivatives is near to this limit and has been measured and found to be 10^{10} for 9,10 dibromoanthracene (3). In several carefully studied cases, the limit can be set at 10^{11} sec^{-1} or even greater. The only notable exception at present is azulene, where the energy separations are anomalously large.

(b) Between the first excited singlet, S_1, and the ground state, S_0, (non-chemical). Rate < 10^6 sec^{-1}. If there is no radiationless conversion to the ground state, it will be true that "The sum of fluorescence and triplet yields, in the absence of chemical change, is unity." This rule is found to be valid in a wide variety of molecules which have been carefully studied, some of them with quite low lying lowest singlet levels (for example, chlorophylls) (14). Since few molecules of interest have lower excited states than chlorophylls, we may expect this to be a general rule. There are many apparent exceptions, some of which show neither triplet formation nor fluorescence, but many of these are known to involve complexities such as dimer formation or other chemical reaction. It seems possible that all exceptions showing apparent radiationless conversion to the ground state involve chemical change which is either permanent or, more usually, transient and reversible.

(c) Between excited states of different multiplicity. Rate 10^6 to 10^{10} sec^{-1}. Although this rate depends on energy separation and on factors which affect the electronic transition probability as, for example, in heavy atoms, most such conversions (inter-system crossings) fall in this range of rates.

(d) Between triplet and ground state. Rate 10^4 to 10^{-2} sec^{-1}. Rates less than 10^3 sec^{-1} are rarely of practical importance in fluid media, being greatly exceeded by the rates of other processes such as quenching by other solutes, particularly oxygen. Diffusion controlled quenching processes, in fluid solvents such as water, have specific rates of 10^9-10^{10} $\ell \cdot mole^{-1}$ sec^{-1}, so 1 $\mu mole/\ell$ of quenching impurity will give a quenching rate exceeding that of radiationless conversion from T_1.

Intermolecular Energy Transfer

In addition to the intramolecular physical processes discussed above, there are numerous intermolecular physical processes in the excited state, of which we will mention only the most important one. The transfer of electronic excitation from one molecule to another similar one, or one with a lower energy level, is of the greatest significance in photobiology where we are usually concerned with a variety of molecules in close proximity. It must be considered whenever an excited molecule is within 100 Å of another molecule of equal or lower excitation energy. This is the average separation in a solution of concentration 2 X 10^{-3} M. The actual transfer distance, R_c, is given by the Förster expression:

$$R_c^6 = \frac{9000\ \ell n\ 10 K^2 \phi_f}{128\pi^6 n^4 N} \int_0^\infty \frac{f_D \varepsilon_A}{\nu^4}\, d\nu$$

where the term after the integral sign is the overlap of the normalized emission spectrum of the excited donor with the absorption spectrum of the acceptor. R_c is the distance at which energy transfer is equal in probability to radiation by the donor so that, if we are concerned with the probability of transfer, the radiative transition probability of the donor cancels out. Thus, for example, the probability of transfer from a triplet and singlet donor, both to give singlet excitation of the acceptor, will be similar if other factors are similar and there are no other deactivation processes operative.

At very close separations of the donor and acceptor molecules, the Forster expression for purely dipole-dipole resonance transfer gives values of transfer probability which are too low, since other interactions become operative. In the limiting case of a crystal or a very concentrated solution, the transfer rate is so great that the excitation can no longer be regarded as associated with any particular molecule at a particular time and one speaks of an exciton process, though it is still possible to discuss such processes kinetically in terms of random walk of the excitation.

In highly condensed systems the excitation may traverse many molecules during its lifetime, and if one of these has a lower lying state, the excitation becomes trapped. A very high degree of purity is necessary to exclude such foreign molecules, and for this reason luminescence is rarely observed from the host molecules in a crystal or highly concentrated solution.

Both singlet and triplet excitation may be transferred at close molecular separations, and their relative importance in such processes as photosynthesis has been discussed. It should, however, be noted that, even if triplet excitation transfer is of higher probability than singlet excitation transfer, this will probably only be the case at intermolecular separations so small that the transfer rate greatly exceeds the singlet lifetime, so that the singlet will usually have been transferred and trapped before intersystem crossing to give a triplet can occur. This seems to be the case in model systems closely related to the chloroplast (5).

PRIMARY PHOTOCHEMICAL PROCESSES

In principle, chemical reaction may occur at any stage during the physical transformations which have been discussed. It is, of course, important to define the electronic state from which reaction occurs; but a further distinction should be made (6) between those reactions which occur in thermally equilibrated vibrational levels of an electronically excited state, which we may describe as equilibrium or E reactions, and those reactions which occur from vibrationally excited levels, either of the ground state or of an electronically excited state, before vibrational relaxation has occurred (See Fig. 3). This latter type of level may be reached either by direct absorption followed by prompt reaction (P) or by radiationless crossing and conversion to a vibrationally excited isoenergetic, lower, electronic level (conversion or C reactions). The most important positions from which the various types of reaction can occur are indicated, with a simple nomenclature by which they may be described, in Fig. 4.

Other processes can readily be accommodated including

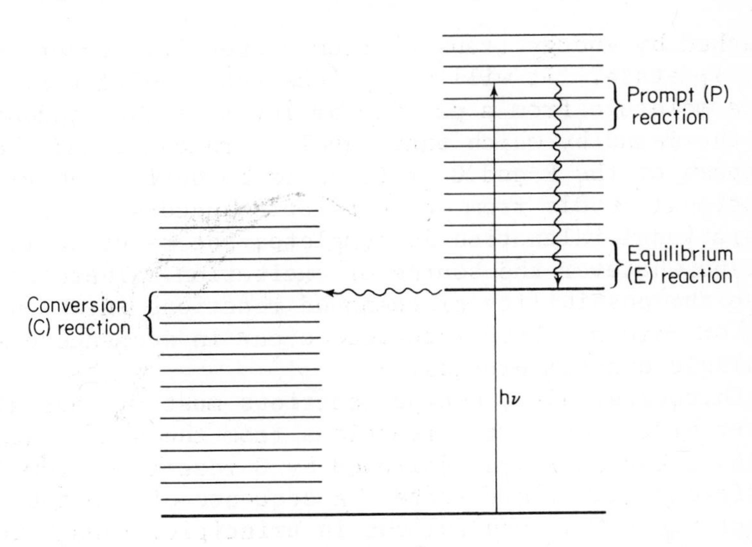

Figure 3

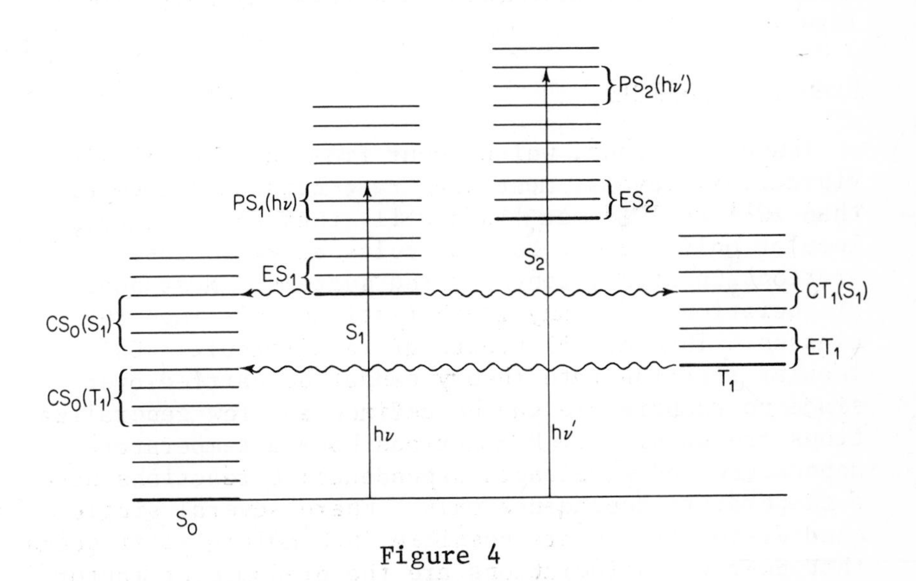

Figure 4

those involving other types of excitation; for example, $PT_1(e)$ signifies a prompt reaction from the initial vibrationally excited first triplet level reached by electron excitation, and PS_1 (donor T_1) signifies prompt reaction of a molecule from a particular level of S_1

reached by energy transfer from a specified donor in its T_1 state. It will be obvious that the reactivity of a molecule from a particular level is independent of the means by which that level is reached, and the purpose of the P and C prefixes is to define the vibrational levels from which reaction occurs. Once vibrational relaxation is complete, the reaction is E-type whatever the source of excitation. There is also the possibility of compound reactions in which two or more of these reactions occur in sequence during a single quantum process.

Chronologically, P-type reactions must precede all other processes, and E reactions from the same state then become possible, followed by C reactions from a different electronic state. A sequence of E and C reactive states then follows in principle, though in practice these possibilities are few in number unless excitation occurs primarily to a very high electronic level.

P and C Reactions

These reactions, which occur from non-equilibrium vibrational levels, must have rate constants greater than 10^{11} to 10^{13} sec^{-1} and will, therefore, be unimolecular only, except for molecules at very close separations in highly concentrated systems. Most bond dissociations and many isomerizations fall into these classes. Theoretical treatment is difficult. Unimolecular reaction rate theory cannot be carried over since no temperature can be defined and few generalizations are possible. P reactions have a temperature dependence and wavelength dependence; C reactions have a temperature dependence only. Where several single bond dissociations are possible in a molecule, it seems that energy considerations are the predominant factor, and the weakest bond is broken with the highest probability. This statement may be modified in solution by operation of the cage effect. Since the reactions are, by definition, dependent on vibrational excitation, it is obvious that energetic considerations will be of prime importance in all reactions of this type. Unless the frequency factors differ widely, we may expect the isomerizations and other reactions of lowest transition state energy to occur with the highest rate. In CS_0 reactions the primary processes should be quite similar to those occurring in thermal reactions, in spite of the fact that the excitation profile is not Boltzmann but nearer to a delta function. This must not be taken to indicate, however, that the final products will be the same, since the conditions subsequent to the primary change are greatly different. An isomer formed in a photochemical reaction of class C is surrounded by molecules at the normal ambient temperature and will itself be quenched to this temperature in less than 10^{-11} sec. The same isomer formed as the initial step in a thermal reaction is surrounded by molecules at the temperature which made the reaction possible, and unless it is more thermally stable than the parent molecule, it will not survive but will undergo further change. It seems probable that many of the remarkable transformations which occur by reactions of type C also occur in the initial steps of thermal reactions but cannot be isolated owing to further thermal transformations.

Although degradative photobiological reactions may proceed by P or C mechanisms, most of the more specific reactions in photobiology, electron and proton transfer for example, are surely of class E. For these reasons, and because more useful generalizations can be made, the rest of this paper will be confined to class E reactions.

Class E Reactions

This is the class of reaction which is of greatest interest to a photochemist because here one is concerned purely with the difference in properties brought about by electronic excitation without the complexities of ill-defined temperature, wavelength dependence, and so forth.

In spite of the variety of chemical structure and reaction, certain generalizations can be made about the relative reactivities of molecules in different electronic states. The factors which differ from one state to another and which may influence reactivity are: (a) Energy, (b) multiplicity, (c) electron density distribution, and (d) orbital symmetry. These will be discussed in turn.

(a) Energy. From the mere fact that, on excitation, an electron in the molecule is raised to an orbital of higher energy, we can immediately conclude that:

(1) The electron affinity in an excited state is higher than that of the ground state.
(2) The ionization potential in the excited state is lower than that of the ground state.

If excitation is from the highest occupied to the lowest unoccupied orbital, the difference in these quantities for the two states is simply the excitation energy.

From (1) and (2) it might at first appear that electron transfer reactions to or from an excited molecule to a second molecule must always occur more readily than to or from the ground state, provided entropy changes are similar in reactions of the two states, which is usually the case. However, in a complex molecule one must take account also of the electron distribution within the molecule, and it is possible that the electron affinity at the reactive site may be reduced on excitation even though the electron affinity of the molecule as a whole increases. In this case, although the energetics of the overall electron transfer reaction may be more favorable in the excited state, the activation energy may be greater and so the reaction rate may be reduced. Examples of this occur with substituted aromatic ketones and quinones where strong charge transfer character in a transition may reduce the rate of transfer of an electron to the carbonyl group (7). Hydrogen atom transfer reactions seem to behave as electron transfer coupled with proton transfer, and so these generalizations apply to them as well.

(b) Multiplicity. The electron spin influences chemical reactivity *per se* only indirectly. Owing to electron correlation, the two electrons in singly occupied orbitals are further separated, on the average in the triplet rather than in the equivalent singlet state; electron repulsion is therefore less, and the energy of the triplet is lower. The two most important types of reaction in the photochemistry of complex molecules are ES_1 and ET_1 reactions. The following general comparisons can be made between them:

(1) Although the reactivity (i.e., the specific rate of reaction per unit time) may be similar in the two states, ET_1 reactions tend to be more common owing to the much longer lifetime of the triplet state (i.e., the lower rate of competing physical processes).
(2) The electron correlation considerations already referred to result in other general differences apart from the purely energetic one. The mixing of charge transfer configurations (i.e., resonance structures) in the triplet state is usually less than in the equivalent singlet. This effect is seen most clearly in proton transfer reactions which are very sensitive to charge density at the deprotonated site in the molecule. It has been found, so far without exception, that the acidity constant of a triplet T_1 state of a substituted aromatic molecule lies intermediate between that of the ground state and the S_1 state (8).

The triplet state is sometimes referred to as a biradical. This is not justified since, in a true biradical, singlet and triplet states would be indistinguishable and there would be no interaction at all between the two electrons concerned. However, since there is less electron interaction in the triplet state, it is true to say that the triplet is more biradical in character than the equivalent singlet.

(c) Electron density distribution. Chemical reactions are primarily changes in electron density distribution within and between molecules; the nuclei follow at a later stage. Since this electron distribution is precisely what is changed on electronic excitation, it is clear that the excited state reached in a photochemical reaction is a new chemical species, not merely a hot molecule. Furthermore, although the energy of the state is significant, it does not tell us much about the reactivity. It is essential to know, as well, something of the type of electron excitation and the electron density distrubitions in the orbitals populated and depopulated during the transition.

For many purposes, a simple classification of orbitals (9) into σ, σ^*, π, π^*, and n (nonbonding) and the corresponding excited states into π-π^*, n-π^*, etc., gives an indication of the change in electron distribution which is very valuable and adequate to predict qualitatively the relative changes in reactivity. With more complex molecules containing two or more electron donating (e.g., NH_2, OH) or withdrawing (e.g., C=O, NO_2) substituent groups, the degree of charge transfer between these groups and the aromatic system is of prime importance and, where this becomes of more importance than the ring π-π^* excitation, the transition and state resulting from it is sometimes described as charge transfer (C-T). Fortunately, the change in dipole moment in the excited state which results from transitions with charge-transfer character is easily estimated from the wavelength shift of the transition in solvents of varying dielectric constant.

Since bimolecular reactions are almost universally of ES_1 or ET_1 type, it is important to know the orbital characteristics of the lowest singlet and triplet excited states. It may happen that two states of quite different orbital type (e.g., n-π^* and π-π^*) lie quite close together and that the order of their energies can be reversed by substituents or by solvent shifts. This can lead to sudden profound changes in reactivity on introducing quite minor modifications of the molecule or its environment which at first seem inexplicable, but on further study appear as beautiful illustrations of many of the principles outlined here. The hydrogen abstraction reaction by substituted aromatic carbonyl compounds, ET_1, and its dependence on solvent, provide a satisfying example of the importance of electron density distributions and orbital type on a particular class of excited state reaction (7).

(d) Orbital symmetry. The concept of conservation of orbital symmetry in a chemical change, introduced by Hoffmann and Woodward (10), is of great importance in any discussion of differences between the reactivity of different electronic states which may have different orbital symmetries. If there are symmetry restrictions on a thermal reaction, it is common to find that those restrictions are reversed in the first excited state leading to a number of subtle distinctions between thermal and photochemical reactions of a concerted type.

Suppose, in a certain reaction, it is necessary to break one or more bonds and to make one or more new bonds. Clearly, if the reaction goes in two steps involving the bond fission processes as the first step, it will have a very large activation energy, at least equal to the energies of the bonds broken. In a concerted reaction the bond making and breaking processes occur simultaneously; in this way it is possible for the reaction to proceed with much smaller activation energy. For such a process to be possible, the bonding character of all the occupied molecular orbitals must be preserved throughout the reaction, and this is only possible if the process is symmetry-allowed. Instead of considering every orbital in the molecule, it is usually sufficient to focus attention on the highest occupied molecular orbital and enquire whether, as the reaction begins along its path, the energy of this orbital rises or falls. Since photochemical excitation invariably changes this very orbital, it clearly affects the symmetry restrictions quite vitally.

Following Hoffmann and Woodward (10), consider the isomerization of cyclobutene into butadiene. If the cyclobutene is substituted with a group at both the 1 and 4 positions, then the stereochemistry of the product butadiene gives a clear indication of the manner in which the bonds rotated during the isomerization reaction. It is found that only one of the two possible stereoisomers occurs, and they are opposite in thermal and photochemical reactions respectively.

If we consider only the highest occupied orbital in ground state butadiene (Fig. 5), it is easy to see that

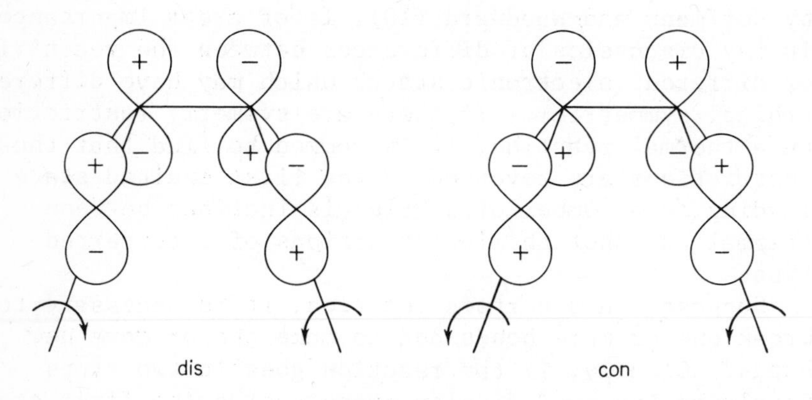

Figure 5

rotation of the orbitals on the 1 and 4 C atoms, where the new bond is to appear, will result in bonding overlap if the rotation is in the same direction (conrotation), but antibonding overlap if the rotation is in opposite directions (disrotation). The disrotatory process will, therefore, have to proceed by a path which has a considerable energy barrier, and in fact the isomeric transformation corresponds to a conrotatory process. Conversely, in the first excited state, the highest occupied orbital is the π^* orbital, and now only a disrotatory process results in bonding overlap throughout the reaction. The situation can also be represented more completely, though perhaps less descriptively, by means of a level correlation diagram (See Fig. 6) as has been done by Longuet-Higgins and Abrahamson. In thermal reactions the highest occupied level in each molecule will be the second, and in photochemical reactions it will be the third. Once again

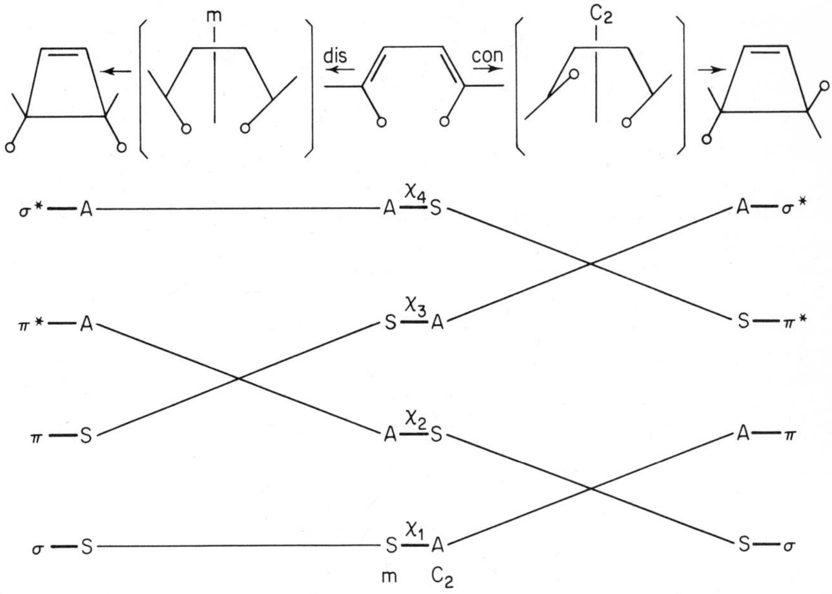

Figure 6

we see how the disrotatory and conrotatory mechanisms are reversed. Other instances are given in reference 10.

Another example of the importance of orbital symmetry considerations in photochemistry is the isomerization of benzene. A detailed prediction is possible of which states of benzene have the various isomerization routes open to them. These predictions have been made by Woodward and Hoffmann and independently by Longuet-Higgins and Bryce-Smith. They are given in Fig. 7 and are in accord with the existing experimental evidence. The stability of Dewar benzene, in spite of its enthalpy of 70 kcal in excess of benzene, is attributed principally to this orbital symmetry restriction on isomerization by a concerted mechanism involving only ground states.

McCapra (11) has recently given an example of the operation of orbital symmetry conservation in bioluminescence reactions. These reactions seem to occur quite commonly by the process shown at the top of Fig. 8.

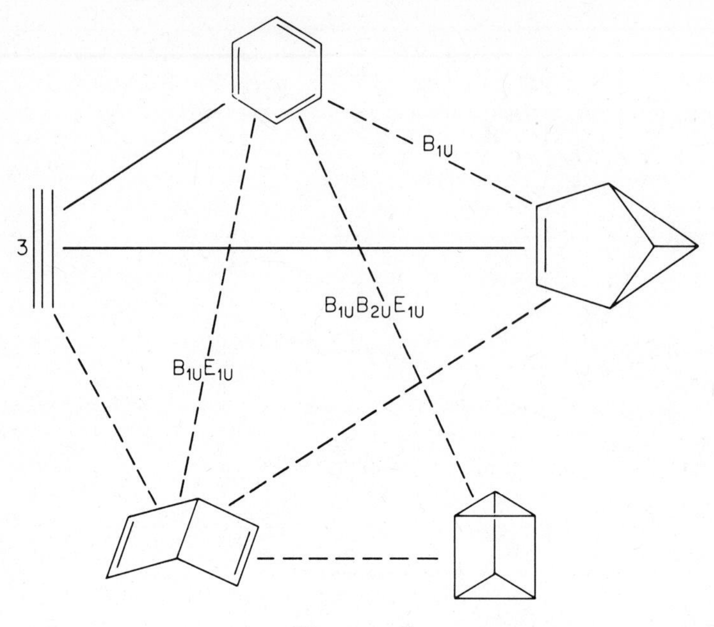

Figure 7

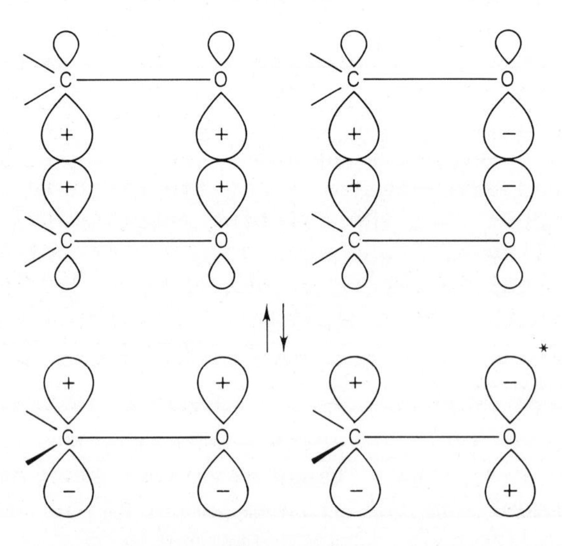

Figure 8

Energetically this presents no problem, since even the formation of the excited state is exothermic. But the question arises why the excited state rather than two ground state molecules should be formed. Again, the formation of two carbonyl ground state molecules is symmetry-forbidden, while the formation of one molecule in an excited state preserves orbital symmetry. Undoubtedly many other applications of orbital symmetry conservation will be found, and they are likely to be of increasing importance in photochemistry.

All these generalizations are equally relevant in both photochemistry and photobiology, and an understanding of the complexities of photobiology must depend heavily on understanding the basic photochemistry.

But it is by no means a one way traffic of ideas. Just as organic chemistry has reached its present successful state by continually enquiring into the nature of biological substances, so we may expect a great acceleration of the science of photochemistry by enquiring how biological systems, with their millions of years experience, have found the most effective way of using the photon.

REFERENCES

1. Rentzepis, P. M., *Chem. Phys. Letters*, *2*, 117 (1968).
2. Robinson, G. Wilse, and R. P. Frosch, *J. Chem. Soc* ., *37*, 1962 (1962).
3. Liu, R. S. H., and J. R. Edman, *J. Amer. Chem. Soc.*, *90*, 213 (1968).
   Liu, R. S. H., and D. M. Gale, *ibid*, p. 1897.
4. Bowers, P. G., and G. Porter, *Proc. Roy. Soc.*, *A296*, 435 (1967), and *Proc. Roy. Soc.*, *A299*, 348 (1967).
5. Porter, G., and G. Strauss, *Proc. Roy. Soc.*, *A295*, 1 (1966).
   Kelly, A. R., and G. Porter, in press.
6. Porter, G., *Nobel Symposium 5*: Fast reactions and primary processes in chemical kinetics. Interscience Publishers, p. 155 (1967).

7. Porter G., and P. Suppan, *Trans. Faraday Soc.*, *61*, 1664 (1965).
8. Porter, G., and G. Jackson, *Proc. Roy. Soc.*, *A260*, 13 (1961).
   Porter, G., and E. Vander Donckt, *Trans. Faraday Soc.*, in press (1968).
9. Kasha, M., *Faraday Soc. Discussions*, *9*, 14 (1950).
10. Hoffmann, R., and R. B. Woodward, *Chem. Soc. Accounts of Chem. Res.*, *1*, 17 (1968).
11. McCapra, F., *Chem. Comm.*, p. 155 (1968).

Human Vision

R. A. Weale

Department of Physiological Optics
Institute of Ophthalmology, London

ON THE PHOTOCHEMISTRY OF VISUAL RECEPTORS

The photochemistry of the receptors as typified by the visual pigments provides the keys whereby quantal carriers of information from the outside world seek admission to the kingdom of our consciousness. Electromagnetic radiation may be absorbed in the retinal rods and cones (Fig. 1), whence the neural message which they may initiate is conducted to the brain *via* other cells and neurons. If I concentrate on the most peripheral link of this chain, namely the photochemistry of the receptor, it is because I believe that a full understanding of peripheral events provides the only basis for a study of more central ones.

Once the photobiologist had only one means available to him for the objective study of the photosensitive contents of an eye: he had to extract them from it.

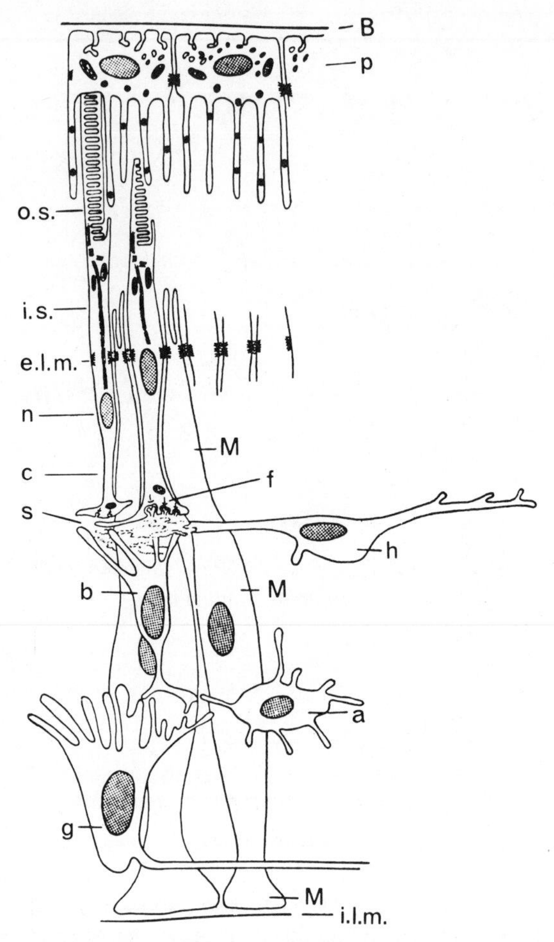

Figure 1. The organization of the vertebrate retina. Light enters from below. B: Bruch's membrane; P: pigment epithelium; o.s.: receptor outer segments or limbs; i.s.: inner segment; e.l.m.: external limiting membrane; n: receptor nucleus; c: receptor fibre; f: receptor foot; s: synapse; b: bipolar cell; h: horizontal cell; a: amacrine cell; g: ganglion cell, the axon of which represents an optic nerve fibre; M: Müller's cell; i.l.m.: internal limiting membrane, abutting the vitreous body. (After Cohen, 1963.)

Following careful purification, he subjected them to spectro-densitometric analysis in a manner familiar to all who have been measuring density spectra for the last hundred years. This task was in no wise facilitated by the extract being light-sensitive, for even if exposure to light did not bleach the so-called visual pigments in their entirety, molecular modifications were brought about. These have been elucidated only in recent years (Hubbard *et al.*, 1965).

But during recent years it has also become possible to study these pigments by methods which appear to leave their environment reasonably intact. I am referring to (a) recordings of the early receptor potential, (b) fundus reflectometry, and (c) monoreceptor densitometry. Shortage of time forces me to confine myself to a discussion of (a) and (b).

What information do we possess on photochemical processes? The molecule of a visual pigment consists of a chromophore, a concatenation of atoms which absorbs visible light, and a protein moiety or opsin, whereby the chromophore is incorporated into the matrix of the retinal receptor. The chromophore (Wald *et al.*, 1963) is chemically related to retinol (vitamin A_1) or 3-dehydro-retinol (vitamin A_2) and is basically a derivative of the respective aldehyde (Morton, 1944).

When visual pigment molecules obtained from a fully dark-adapted retina are exposed to light, they are bleached. In the first instance, their stereochemical configuration changes from the 11-cis (Fig. 2) to the all-trans form, and they may be finally converted, for example, into retinol aldehyde or retinald (Hubbard *et al.*, 1965). At the same time, one or more cross-linkages between the chromophore and its carrier may be broken (Hubbard, 1958). Some workers hold that the visual impulse is generated, if not by actual isomerization of the chromophore, then at least by an event following it in close succession (Wald *et al.*, 1963). Others stress that the opsin released from constraint by the snapping of the linkages may change its configuration and thereby provide a charge displacement which initiates the visual response (Abrahamson *et al.*, 1960). There is a great deal to be said in favor of the latter

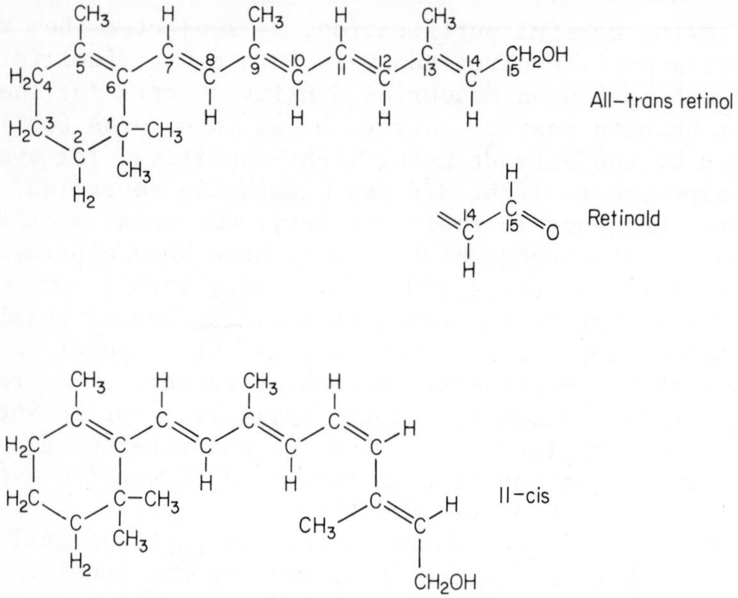

Figure 2. Three configurations of the visual pigment molecule, with the 11-cis form being found in the dark-adapted eye. Bleaching by light involves a conversion to the all-trans form and then to retinald.

view. However, our ignorance of the structure of the opsin, let alone of the changes it might undergo when freed from chromophoric restraint, delays unreserved acceptance of this alternative. For the time being, it may be simpler to consider some of the changes following isomerization and to see what counterparts, if any, can be detected by electric means.

The first product that has so far been detected after rhodopsin has been exposed to light is prelumirhodopsin (P) (Yoshizawa and Wald, 1963), which is stable in solution only at temperatures lower than -180°C. This product is converted into lumirhodopsin by thermal processes (Fig. 3), and this changes in turn into metarhodopsin I (MRI). At body temperatures the lifetime of this intermediate lasts only about 30 μsec; it is in tautomeric equilibrium with metarhodopsin II (MRII) which can change into transient orange

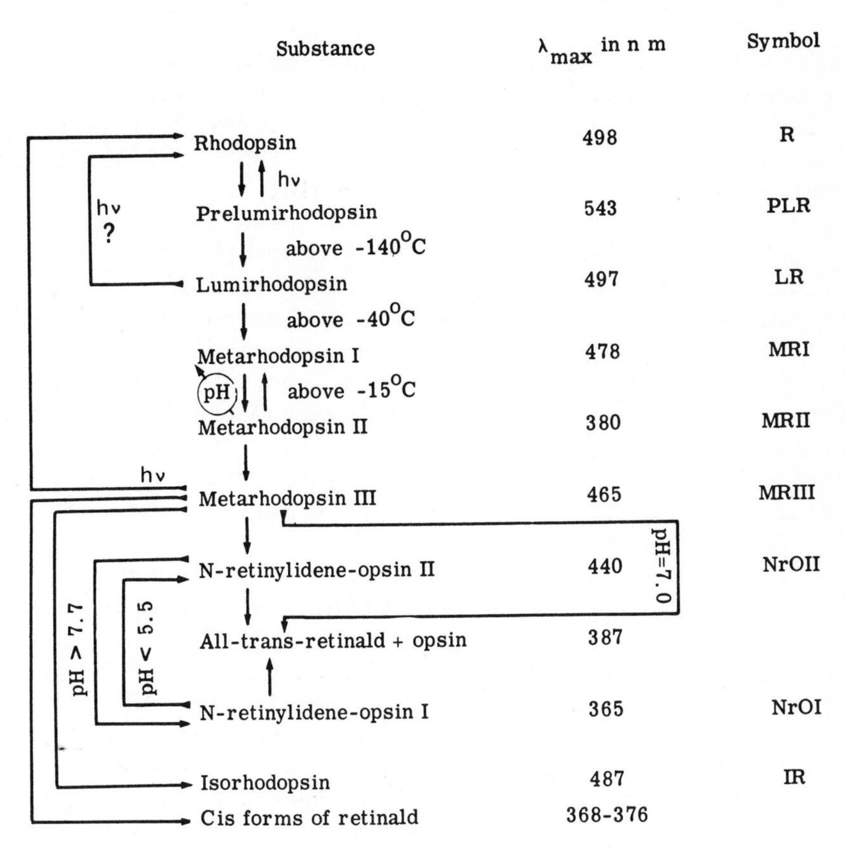

Figure 3. Sequence of conversion events leading to a variety of states of the visual pigment, with absorption maxima indicated for each state.

(TRO or MRIII) under suitable conditions. TRO is relatively stable in the living eye, e.g., that of man (Weale, 1967). The remainder of the events can be gleaned from Fig. 3, but they are most unlikely to have any direct bearing on the elicitation of the visual impulse. We shall, therefore, dismiss them from all further consideration. However, one other point requires immediate attention. When one or the other of the above intermediate substances is illuminated, photic changes may occur. There is, indeed, a finite probability of prelumirhodopsin (P), lumirhodopsin (L), MRI, II, or TRO being converted to the original parent pigment, e.g., visual purple (rhodopsin). As the thermal processes can therefore be counteracted by light, this phenomenon is called photoreversal (Hubbard and Kropf, 1958; Bridges, 1961; Williams, 1964; Villermet and Weale, 1966; Ripps and Weale, 1968; Rushton and Henry, 1968). It is said to occur when the bleaching flash is so intense as to give a chromophore a finite chance to absorb at least two quanta of light within a short interval of time. This stipulation ensures that if the first quantum produces an intermediate substance which can photorevert, the latter may have a finite chance of absorbing the second quantum before decaying to a compound which cannot photorevert to the first (parent) substance. It will be noted that the decay time plays a cardinal role in this matter, and that it varies with the temperature.

Let us now turn our attention to the electrical response of the retina. When one electrode is placed on the cornea and another, said to be indifferent, is placed on the lid or some other nonocular part of the head, a flash of light can be shown to give rise to a change in the standing potential of the eye. This flash of light is called the electroretinogram. It consists of the a-wave, which is negative in relation to the cornea, a positive b-wave, a slower c-wave, and an off-effect or d-wave. In animals the electroretinogram can also be recorded when the active electrode rests on, or passes through, the retina. It is then found that the relative amplitudes of these components vary with the depth of penetration even though the stimulus intensity (which also determines this amplitude) may be kept constant (Brown *et al.*, 1965). With electrode-marking techniques it is possible to release a histological stain into the retinal tissue and to determine the structure wherein a given potential change is most likely to originate. In this manner it was shown that the a-wave is to be associated with the inner limb of the photoreceptors, whereas the b-wave originates in the bipolar layer of the retina (Brown *et al.*, 1965). Earlier studies had shown that the c-wave arises in the iris or in the pigment epithelium, and the d-wave represents a continuation of the a-wave, unmasked by the cessation of the b- and c-waves. Evidently, then, the a-wave represents the earliest potential change arising most peripherally in the visual path; but the fact that its point of origin lies in the inner limb, while light is absorbed in the outer limb, makes it impossible to accept the a-wave as the primary receptor potential.

However, new candidates have recently applied for this post. When very high intensity stimuli are applied to the retina two distinct potential changes are found to precede even the a-wave. Because of this and because there is good evidence to associate them with the outer limb, they are collectively referred to as the early receptor potential (Brown and Murakami, 1964a, b; Cone, 1965; Pak, 1965). Their respective latent periods are virtually zero, but as one of them reaches it peak before the other, they are designated as R1 and R2 respectively (Fig. 4). The former is positive in relation to the cornea, while the latter is negative, like the a-wave.

An outstanding piece of energy calibration enabled Cone (1964) to show that the amplitudes of both components are proportional to the number of pigment molecules bleached. We are therefore dealing with a fundamental graded potential. Insofar as a molecule of pigment is isomerized by a single quantum of light, the mechanism seems to represent a quantum counter (Fig. 5). Perhaps the emphasis on the term "fundamental" should be reinforced. The early receptor potential (ERP) is observed not only in vertebrate receptors (cf., Arden, 1968) but also in invertebrate ones (Brown *et al.*, 1967): in eye-cups containing melanin but no retina and in green leaves (Arden *et al.*, 1966; Crawford *et al.*, 1967; Ebrey and Cone, 1967). That melanin might act as a transducer of electromagnetic energy had been foreshadowed earlier (Weale, 1956). It would seem that the absorption of energy gives rise to an R1-R2 response even in tissues not endowed with specially organized photic receptors. But even if the ERP can be treated as a quantum counter, this conclusion is empirical: it does not explain the nature of the phenomenon. An attempt to do this takes us back to

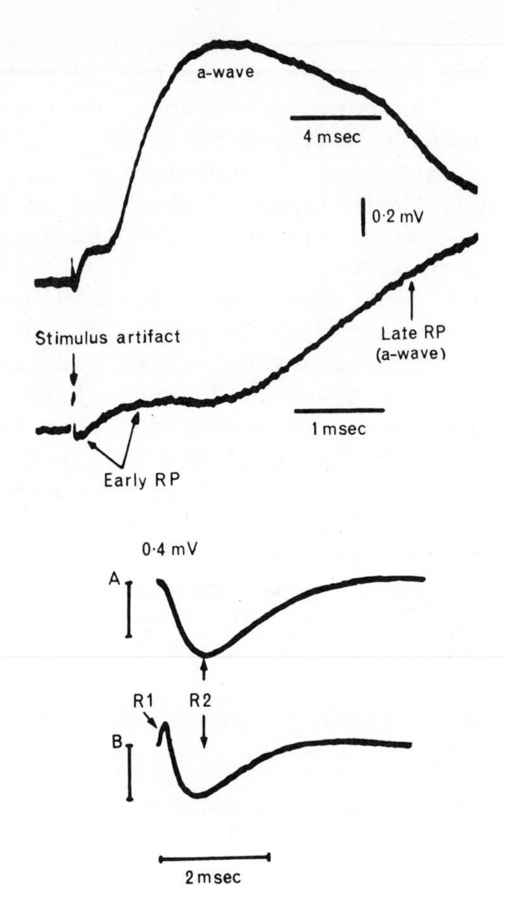

Figure 4. (Top) The early receptor potential (ERP) recorded from a monkey fovea. The upper trace was recorded under conditions identical with those obtaining for the lower trace except that the time of sweep of the cathode-ray electron beam was reduced by a factor of 4 in order to show the relation of the ERP to the a-wave. (After Brown *et al.*, 1965). (Bottom) The ERP in the rat (A) and frog (B). R1 and R2 represent two phases of this potential, which have different parameters. The a-wave, which would normally overlap the decay phase of R2 was suppressed by intense light-adaptation. Polarity: R1, corneal positive. R2, corneal negative. (After Cone, 1965.)

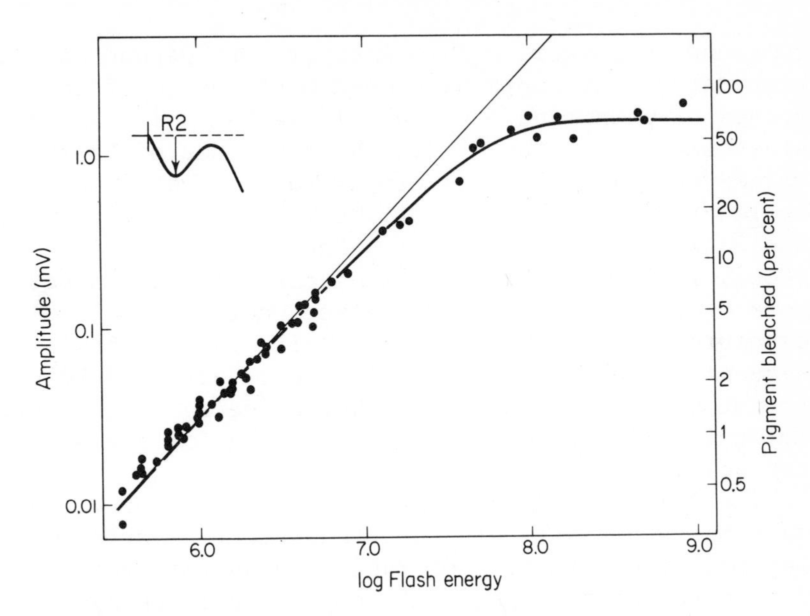

Figure 5. Amplitude of the R2 component of the early RP in the albino rat as a function of flash energy on a log log plot. The flash energy is the energy of the stimulating beam. It is calibrated so that log (flash energy) = 0 is the beam energy at which one quantum is absorbed by the average rod per flash. The narrow straight line has a slope of 1. The broad curved line shows the calculated fraction of pigment bleached by the flash. It is drawn relative to the theoretical scale of pigment bleached shown on the right. Flash duration: 0.7 msec. (Cone, 1964.)

photochemistry.

As the ERP, contrary to its designation, represents a *change* in potential, it is useful to remember that a net change in one or more substances must underlie it. This implies that the formation of a substance has to be considered in relation to its possible disappearance. Products of bleaching are subject to so-called thermal decay, as is implicit from Fig. 3. As the amplitudes of the components of the ERP and the rates at which one compound is converted into another both depend on the temperature, attempts have been made to see if the two groups of phenomena are causally linked. It has to be stressed at once that it is the amplitude of R2 which varies considerably with temperature. When an eye is cooled to 0-5°C, R2 is abolished. It is similarly abolished when the retina is heated to just over *50°C* (Cone, 1965). It is noteworthy that above a temperature of 60°C, rhodopsin can bleach not only following exposure to light but also thermally in the dark. Consequently the failure of R2 in the range of 50-60°C cannot be attributed to the thermal decay of the parent molecule, although later stages may be implicated (but see below).

The fact that the amplitude of R1 does not depend on temperature suggests that the potential arises during the transformation of rhodopsin and/or prelumirhodopsin to metarhodopsin I. Lumirhodopsin decays too fast at body temperature to be able to offer a likely alternative (cf., Pak and Boes, 1967), and the balance of the evidence favors either the formation or the decay of prelumirhodopsin as the factor generating R1.

While the literature shows agreement on the origin of R1 occurring at a stage prior to the formation of MRII, there is no unanimity as regards the formation of R2. There is some evidence for the view that it is generated during the change of MRI to MRII. Thus Cone (1967b) has shown that the wave-form of R2 is a mirror image of that of the potential (ꓤ2) recorded when MRII is changed by photoreversal to rhodopsin (See Fig. 6); also R2 is abolished at low temperatures as would be expected on the basis of the photostability of MRI which is observed below 0°C (Matthews *et al.*, 1963/4). However, monovalent cations, such as isotonic RbCl, NH_4Cl, or KCl, enhance the amplitude of R2 without affecting R1 (Pak *et al.*, 1967). As both potentials have been related quantitatively to the amount of pigment bleached (Cone, 1965), provided this does not exceed 30% of the amount present in the dark-adapted eye (Arden *et al.*, 1966a), such an enhancement suggests that the generation of R1 and R2 may depend on different processes. This view is shared by Arden *et al.* (1966b),

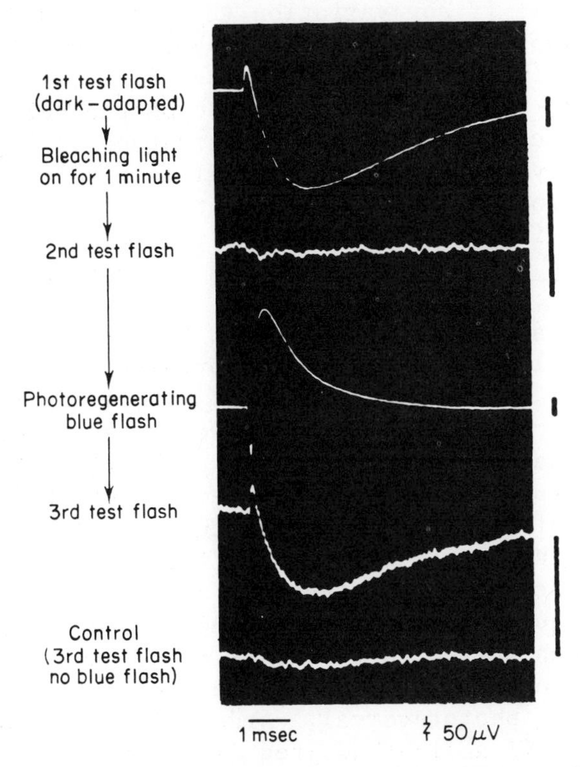

Figure 6. Photoregeneration of the early RP in the eye of the albino rat. Both the test flash and the bleaching light consisted of long wavelengths primarily absorbed by rhodopsin. The blue photoregenerating flash contained wavelengths absorbed by the longer-lived intermediates of the bleaching process. The control trace was obtained from a second eye subjected to the same bleaching exposure and test flashes, but without the blue flash. Temperature: 27°C. (Copyright 1967 by the American Association for the Advancement of Science.)

Arden (1968), and Arden and Miller (1968) for a number of reasons. Over the pH range of 4 to 8, when the tautomeric equilibrium between MRI and MRII varies considerably (Ostroy *et al.*, 1966), the amplitude of R2 shows little if any variation (Brindley and Gardner-Medwin, 1966). Yet the addition of sodium bisulphite to the retina under test increases R2 (Arden *et al.*, 1968) even though MRII is said to be unobservable in its presence. Again, the time-constant of R2, corrected for any contribution which R1 is likely to make to the wave-form of the later potential (Arden and Miller, 1968), is incompatible with the rate constant of the conversion of MRI to MRII at the appropriate temperature.

Before attempting to pinpoint the origin of R2, it is desirable to round off the correlation between photochemical and electrical events with a consideration of what occurs during the photoregeneration of rhodopsin. We have stressed that the proportionality between the amplitudes of the early receptor potentials and the quantity of visual pigment molecules bleached is confined to the range covering the first third of molecules bleached. Thereafter the amplitude is smaller than the quantal input would lead one to expect, and it has been suggested that this is due to the number of molecules bleached becoming effectively smaller as a result of photoreversal. Arden *et al.* (1966a) have shown that if a light-adapted (i.e., partly bleached) retina is exposed to a series of flashes, progressively larger amplitudes are recorded if the light includes a significant violet component, but that, on the contrary, a diminution occurs when this component is absent. The increase is attributed to photoregeneration, the decrease to plain bleaching. The associated potentials, called photoreversal potentials (Fig. 6), were found by Cone (1967) to have specific action spectra peaking at 480 and 380 nm respectively. It is permissable to deduce from this that the two photoreversal potentials ꓤ1 and ꓤ2, which are mirror-images of R1 and R2, are produced when metarhodopsin I and II are illuminated so as to photorevert to rhodopsin. The polarities of the reversal potentials are opposite to those of the ERP, but the time courses are not strictly parallel presumably because the regenerating chromophores do not accurately retrace their paths to decay. In this context Arden *et al.* (1966a) noted in a densitometric study, which paralleled their electrical investigation, that photoreversal is accompanied by the formation of a pigment presumed to be isorhodopsin. While their conclusion that another pigment is formed in these circumstances is not open to question, their results lead to the query whether isorhodopsin does not absorb maximally perhaps at 478 nm rather than 486 nm (cf., Hubbard, 1965), or alternatively, whether the pigment, the formation of which Arden *et al.* observed, is not really transient orange (TRO). The fact that the pigment remained thermally stable for several days appears to point to the formation of isorhodopsin, but the spectral evidence does not fully support this.

Goldstein (1966) has shown that the action spectra of R1 of both dark- and light-adapted frog retinae obey the spectral function to be expected if the potential were mediated only by cones (cf., Brown and Murakami, 1964). This is surprising since the number of rhodopsin (rod) molecules in the frog retina far exceeds that of cone-pigment molecules even though receptor for receptor, the number of rods may well be of the same order as that of cones (Villermet and Weale, 1965). Why then does a mammalian pure-rod retina like that of the rat produce a relatively large ERP? One possibility, mentioned by Cohen (1968), is based on the peculiar structure of frog rods: the discs from which they are built up appear to be isolated from one another and so prevent the development of requisite membrane potential. Another relevant factor wherein frog rods differ significantly both from mammalian rods and frog cones is their size: the latter receptors are very much smaller and thinner than frog rods. Although these obviously contain much more pigment which appears to be distributed uniformly throughout the receptor outer limb (Wolken, 1963), the ratio (surface area/volume) is 10-50 times smaller in frog rods than in frog cones or mammalian rods. It is thus possible that a given number of pigment molecules can usefully contribute to the production of the early receptor potential only if there is a relatively large membrane area for them to act upon.

Further evidence for the view that the receptor has

to be intact for an ERP to be initiated was obtained from a study with the light microscope.

Receptor outer limbs have been known for about one hundred years to be birefringent and to exhibit marked optical activity. If placed on a microscope stage between a polarizer with its plane of polarization pointing at 12 o'clock, and an analyzer crossing it, then those outer limbs pointing at 10:30 are found to transmit a maximum amount of light. Villermet and Weale (1968) have repeated and extended the classical studies of Schmidt (1938). They showed that the change in birefringence of outer limbs following their immersion in alcohol is more likely to be due to denaturation of the rod protein than to the elution of lipid material (as Schmidt believed). This discovery led them to the study of several fixatives, and of temperature effects. Activity was observed in the 10:30 (Fig. 7a) direction in fresh,rods, and this activity was dominant over subsidiary activity in a direction at approximately

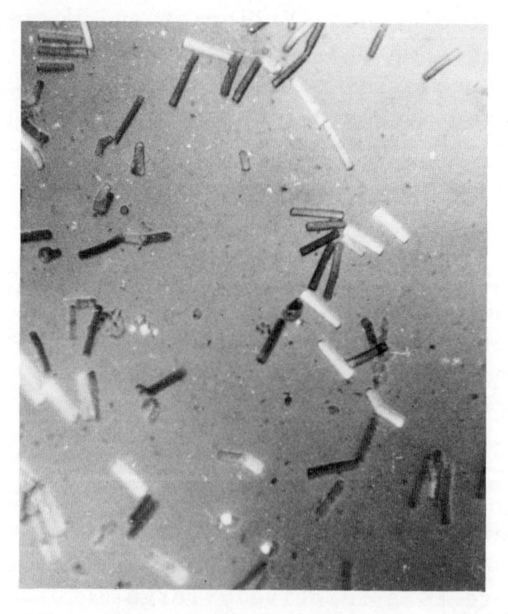

Figure 7a. Frog rods in Ringer. (Villermet and Weale, 1968).

right angles to it. The two activities were therefore designated with 10:30 and 1:30. The following parallels with the behavior of R2 were established.

1. Glutaraldehyde fixative abolishes 10:30 and enhances 1:30 (Fig. 7b); it abolishes R2 (Arden *et al.*, 1968).

Figure 7b. Loose and attached frog rods fixed in glutaraldehyde. (Villermet and Weale, 1968.)

2. Formaldehyde preserves normal activity for hours; it maintains R2 (Arden *et al.*, 1968).
3. Cooling decreases 1:30 in relation to 10:30, it abolishes R2.
4. Heating for 2 min in Ringer at 51°C abolishes 10:30 and enhances 1:30; it abolishes R2 (Cone and Brown, 1967).
5. 50% (v/v) glycerol in Ringer enhances 1:30 within 15 min; it abolishes R2 within 50 min (Brindley and Gardner-Medwin, 1966).

6. 40% ethyl alcohol in Ringer is without effect on birefringence but higher concentrations produce denaturation; the same concentration is without effect on R2, but an unspecified higher concentration destroys it (Fatechand, 1968).

Thus R2 appears to be maintained only when the molecular configuration of the rods, as manifested by optical activity, is preserved in its normal state. This is true of fresh rods at "physiological" temperatures and after formaldehyde fixation. These observations lend tentative support to the view that R2 can be produced only if the submicroscopic state of the rod is intact: for example, the temperature effect (4) is particularly striking because, as we previously noted, visual purple becomes subject to thermolysis only at temperatures above 60°C, yet both birefringence and R2 are radically interfered with 9° below this temperature. While this work was being prepared for press, my attention was drawn to the fact that the birefringence of rat rods changes at 58°C (Cone, 1968). It remains to be seen whether the difference between poikilothermic and homothermic rods is significant.

ON HUMAN CONE PIGMENTS

In the second part of this survey, I should like to turn to some aspects of fundus reflectometry with special reference to human cone pigments. In view of the data obtained by Brown and Wald (1964) and others, fundus reflectometry may seem old hat. But it does help to silhouette some important problems which bear not only on the apparent concentration of the pigments in the receptors, but also on their spectral location. It will be recalled that the technique is a type of objective ophthalmoscopic examination of the eye. A monochromatic test-beam is sent into the eye, traverses the retina, is partly reflected at the back of the eye, retraverses the retina, and finally emerges from the eye (Fig. 8). Its intensity can now be measured (e.g., photoelectrically) and will clearly depend on the

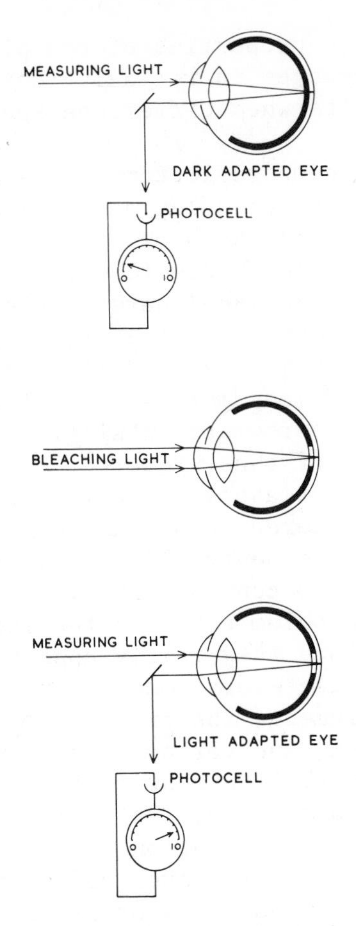

Figure 8. The principle underlying fundus reflectometry. (The meter reading is not to scale.)

amount of absorption it has suffered along its path. Other factors being equal, such absorption varies with the nature and concentration of the pigments traversed by the test-beam. Their nature is determined genetically, but the concentration can be reduced by the intensity of a bleaching beam. If the wavelength of the test-beam is varied and its intensity measured (e.g., before and after a given retinal area has been exposed to a bleaching beam), different test-wavelengths will record different changes in absorbance, depending on the absorption properties of the pigment(s) in question; these changes in retinal transmissivity can be in density units when difference spectra can be plotted.

Although fundus reflectometry has been used by Stiles (1939) and Brindley and Willmer (1952), only Carr *et al.* (1966), Rushton (1958; 1963a, b, c; 1965a, b, c, d), Ripps and Weale (1963, 1964, 1965a, b), and Weale (1959, 1965, 1968) have studied human cone pigments *in vivo*. Whatever the true value of the pigment density *in situ* may be, and it is subject to several corrections and many more doubts, the density values obtained by measurements in single receptors are a small fraction of those recorded by fundus reflectometry. Possible reasons for this serious discrepancy have been considered (Ripps and Weale, 1965a; Weale, 1968; Rushton and Henry, 1968), but they do not seem to be specific to cone pigments.

The basic problem concerns the isolation of these substances, since they rarely appear in a retina one at a time. While every cone presumably contains only one type of pigment, the primate fovea contains a selection of them, and various attempts have been made to sift the mixture. The most radical method was employed by Rushton (1963a, b, c; 1965a, b). Making the eminently reasonable assumption that one of the respects wherein color-defective eyes differ from normal ones is in the types or numbers of pigment contained in their fovea, he made a detailed study of protanopes (red-defective) and deuteranopes (green-defective). One of the tests for the homogeneity of an assembly of pigment molecules consists in exposing the photolabile material to various chromatic bleaching lights. Then the shape of the density difference spectrum will be found to remain constant only when one type of pigment is present, whereas it will change in a predictable manner if two or more coexist.

In the case of the protanope, Rushton obtained a difference spectrum that is unusual, as stressed by Brown and Wald (1964), in that it is flat-topped and contains no datum at wavelengths shorter than 520 nm. An attempt to correct for stray light led Rushton to the view that the pigment absorbs more than 50% of the incident light. This is a high value as compared with the 20-30% (e.g., of rhodopsin) typically contained in human rods. Even though it clearly accentuates the discrepancy between *in vivo* measurements and the few per cent absorption observed in monoreceptor densitometry, this observation would be exceedingly valuable inasmuch as a number of subjective experiments (Enoch and Stiles, 1959; Walraven and Bouman, 1960) demand pigment densities which, for the wavelength of maximum absorption, should amount to almost 1.0. Unfortunately such a high density is incompatible with the shape of the spectral sensitivity curve of protanopes which the difference spectrum might be expected to agree with (Fig. 9) if certain corrections are applied (Weale,

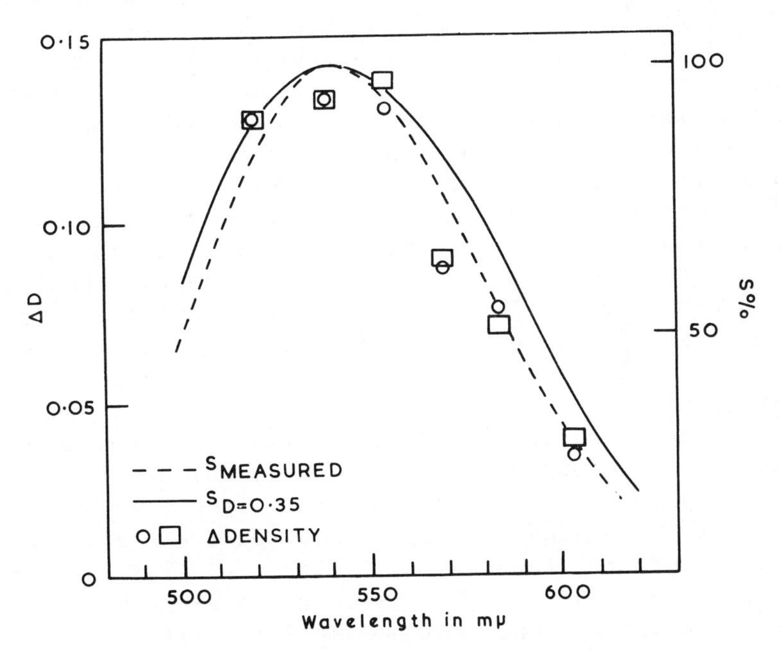

Figure 9. Symbols of Rushton's measurements. Dashed curve: protanopic sensitivity function. Continuous curve: what this function would be expected to look like if Rushton's computed density maximum of 0.35 were valid. (Weale, 1964.)

1964). The band-width of an absorption (and action) spectrum increases with the maximum density, but such broadening is not observed in data obtained on protanopes. This fact renders Rushton's theoretical approach to his results open to question. Nonetheless, the data strongly suggests that the principal pigment of the protanope peaks in the neighborhood of 540 nm and that it lacks a red-sensitive pigment.

As the spectral sensitivity curve of deuteranopes (Hsia and Graham, 1957) differs significantly from that of the protanope, it is natural to expect a difference in the pigment content of their fovea. Originally Rushton (1958) believed that the deuteranopic cones contain two unsegregated pigments. One of these was the 540 nm pigment found in the protanopes, and the other, a red-sensitive pigment absorbing maximally at 590 nm. The spectral position of the latter tended to oscillate in later publications, a circumstance which Ripps and Weale (1964) attributed to variations in the intensity of the red bleaching light employed, and which Rushton (1965a) attempted to explain in terms of a stray light effect. It is hard to share his sense of the importance of this factor because he showed in an elegant experiment (1965c) that stray light does not amount to more than 6%. Even this percentage includes a moiety which does not vary with the wavelength (cf., Weale, 1966). In order to circumvent the alleged untoward effect of stray light on the difference spectrum, Rushton expresses his observations in terms of fractional transmission differences. If the transmissivity of the pigment is T_1 at one level of bleaching, T_2 at another, and T_0 when all the pigment is bleached away, then the density difference spectrum is replaced by the spectral function of $f(T) = (T_1 - T_2)/T_0$. It is readily seen that this is equal to $\exp(-kD_1) - \exp(-kD_2)/\exp(-kD_0)$ which simplifies with little error to $D_2 - D_1 = \Delta D$ when $\Delta D < 0.3$. This explains why the function successfully matches the deuteranopic sensitivity curve (Fig. 10) in a manner analogous to that achieved with orthodox density spectra. Rushton concluded from this figure (and other data) that deuteranopic fovea possess only one pigment (other than per-

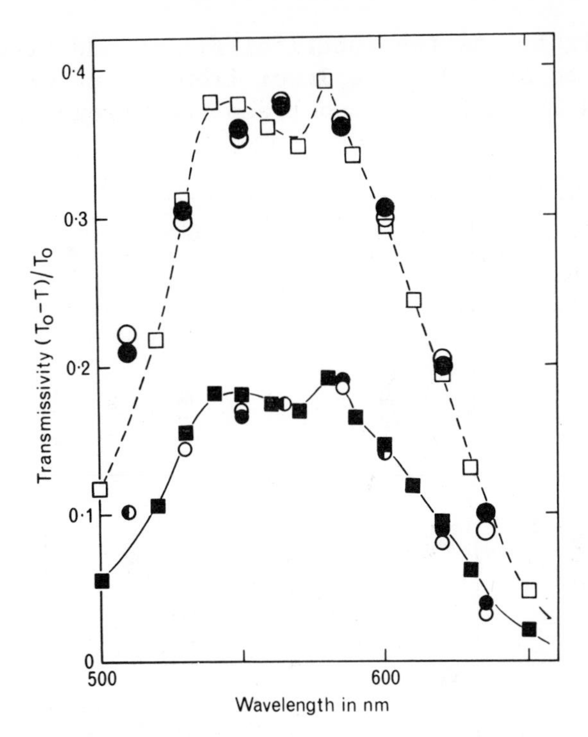

Figure 10. Small black circles (●): relative transmissivity differences when a deuteranopic fovea was bleached with red light. Small white circles (○): for a deuteranopic fovea bleached with blue-green light. Large circles: corresponding data when the bleaches were completed with white light. Black and white squares: scaled values of Hsia and Graham's deuteranopic visibility function. (After Rushton, 1965a.)

haps one absorbing mainly in the violet part of the spectrum). The absorption of this pigment is maximal in the neighborhood of 570 nm. However, the aforementioned figure deserves a closer look. The lower data represent the change in transmissivity following a bleach with deep red light (●) and, in a separate experiment, results for glue-green light (○). The upper data show corresponding changes in transmissivity when the partial chromatic bleaches were completed with white light. As the function f(T) is additive, we are allowed to subtract the lower from the upper data (Fig. 11), and the difference between the strong white bleach

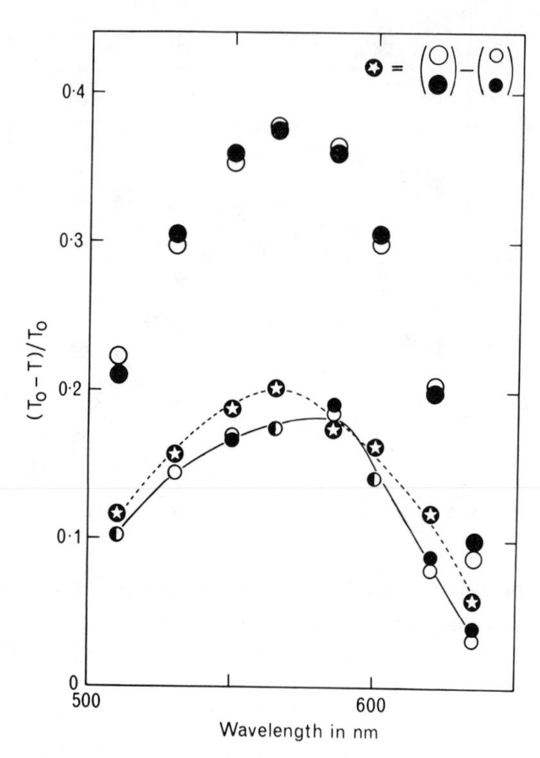

Figure 11. Circles: see Figure 10 for their meanings. Asterisks: the difference between the two sets of data.

and the spectral bleaches, which peak at about 560 nm, clearly calls into question Rushton's conclusion that the deuteranopic fovea possesses only one pigment absorbing in the central part of the spectrum (at 585 nm). Ripps and Weale (1965a) have shown that there are situations when red and green bleaches done at but one intensity level may fail to demonstrate the presence of more than one pigment. And while Wald and Brown's (1965) recent studies support the view that the deuteranope has only one "central" pigment, just as Pitt's

(1936) and Willmer's (1949) data suggest that it has two, the evidence obtained from fundus reflectometry is so far inconclusive.

The analysis of the color-normal fovea is obviously even more complicated. For a start it is hard to make legitimate assumptions regarding the relative distribution of red- and green-sensitive cones even if we assume that blue-sensitive ones are not well represented in this retinal region. It is, moreover, virtually impossible to bleach only one of the pigments to the exclusion of the others. If, nonetheless, a relatively feeble deep red light is used for a limited period of time, a difference spectrum can be obtained which shows clear dominance in the orange part of the spectrum. This is shown in Fig. 12 with the data of Rushton

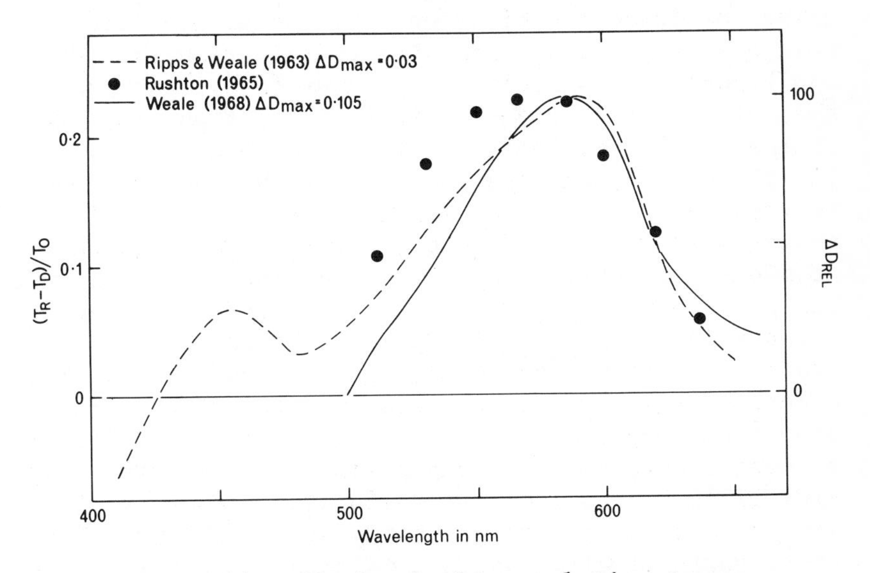

Figure 12. Black circles: relative transmissivity differences following a red bleach of the normal fovea (Rushton, 1965d). Dashed curve: average density difference spectrum for two normal foveae (Ripps and Weale, 1963). Continuous curve: see text for details (cf. Fig. 13).

(1965d) and Ripps and Weale (1963). On the hypothesis that red-sensitive cones are concentrated in the center of the fovea at the expense of green-sensitive ones, Weale (1968) measured difference spectra with two test-fields: one sampled only the central circular area whose diameter at the nodal point subtended an angle of a little over 20 ft, the other sampled a larger rectangular area (80 ft X 40 ft). The averaged data for two subjects gave rise to the two functions (S and L) shown in Fig. 13. Assuming that these functions differ only as regards the relative contributions of green- and red-sensitive pigments, Weale selected Wald and Brown's spectral function for a green-sensitive cone as determined by monoreceptor densitometry and computed the function corresponding to the red-sensitive moiety (Fig. 13). The two difference spectra could be described by these functions only if the small-field data were assumed to have a smaller contribution from Wald and Brown's green-sensitive pigment (i.e., perhaps a smaller number of cones containing it) than was true of the large-field data. This may provide a substrate for the well-known observation that the central fovea is very sensitive to red light. The value of this analysis is restricted (e.g., all the density change at 500 nm is attributed to the green-sensitive pigment), but our ignorance of the effects of products of bleaching possibly absorbing in this region would make any other assumption equally arbitrary. Nonetheless, in spectral regions where the contribution of the green-sensitive component is relatively feebler ($\lambda > 550$ nm), agreement between this computation and the curve from Fig. 11 is suggestive. It would appear that the red-sensitive pigment acts maximally at 580-590 nm. This is at variance with conclusions drawn from work on isolated receptors which puts the maximum at 565 nm.

We see, therefore, that these objective methods whereby the basis underlying mechanisms of color vision are being probed agree poorly with each other, not only as regards the apparent concentrations of pigments located in cone outer limbs, but also (and this applies more to the red than the green-sensitive pigment) as regards their spectral location. On the other hand

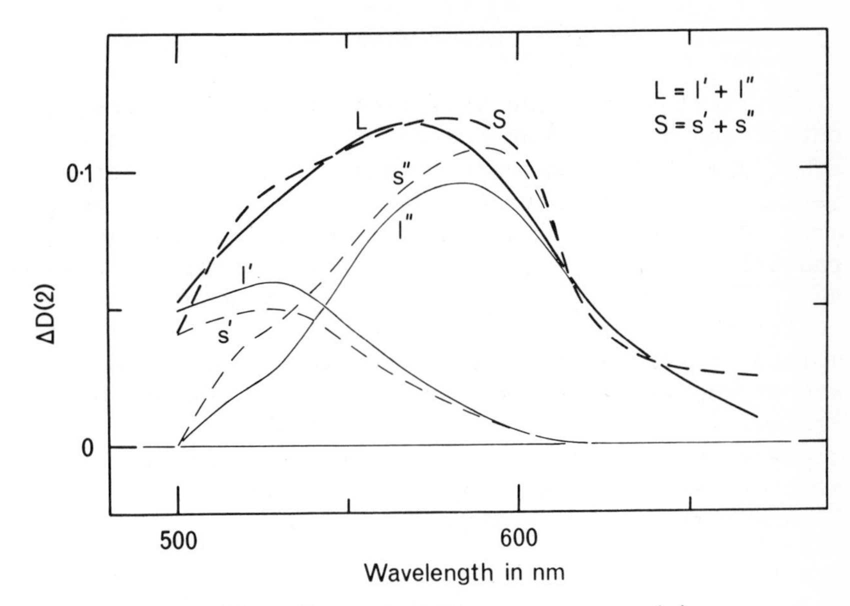

Figure 13. Heavy continuous curve (L): average of difference spectra obtained for two normal foveae bleached with deep red light [(6.51 log phot td) (30s)], the measuring field being large. Heavy dashed curve (S): the same for a small measuring field. If these two functions are assumed to be composed of two spectral moieties and Wald and Brown's green-sensitive function is given by the primed symbols, the double-primed functions can be computed. As "s" and "l" agree tolerably well, their average is shown in Fig. 11 (continuous curve).

the recent work shows that products of bleaching can be detected in the central fovea, and this is in agreement with the earlier observations of Brown and Wald on isolated cones. This concord needs stressing in view of the denial by some authors that such products are detectable by methods employed heretofore.

CONCLUSION

At first glance the selection of my two topics, two out of any number that present themselves, might seem arbitrary, but it was not made aimlessly. Rather was it designed to illustrate that even at the most peripheral points of the visual path, where electromagnetic radiation is absorbed and causes a chemical event which again gives rise to an electrical phenomenon, the solution of one problem creates another. Perhaps it has helped to stress that our understanding of the processes at the visual center, which involve the act of seeing as distinct from that of visual stimulation, depends on the replies we can give to a set of unanswered questions that crop up at the initial traffic lights on the road to sight.

REFERENCES

Abrahamson, E. W., J. Marquisee, P. Garuzzi and N. Roubie, *Z. Elektrochem.*, *64*, 177-180, (1960).
Arden, G. B., in press, (1968).
Arden, G. B., Ikeda Hisako and I. M. Siegel, *Vision Res.*, *6*, 357-371, (1966a).
Arden, G. B., Ikeda Hisako and I. M. Siegel, *Vision Res.*, *6*, 373-384, (1966b).
Arden, G. B., C. D. B. Bridges, Ikeda Hisako and I. M. Siegel, *Nature*, *212*, 1235-1236, (1966).
Arden, G. B., C. D. B. Bridges, Ikeda Hisako and I. M. Siegel, *Vision Res.*, *8*, 3-24, (1968).
Arden, G. B. and G. L. Miller, *Nature*, *218*, 646-649, (1969).
Bridges, C. D. B., *Biochem. J.*, *79*, 135-143, (1961).
Brindley, G. S. and A. R. Gardner-Medwin, *J. Physiol.*, *182*, 185-194, (1966).
Brindley, G. S. and E. N. Willmer, *J. Physiol.*, *116*, 350-356, (1952).
Brown, J. E., J. R. Murray and F. G. Smith, *Science*, *158*, 665-666, (1967).
Brown, K. T. and M. Murakami, *Nature*, *201*, 626-628, (1964a).

Brown, K. T. and M. Murakami, *Nature*, *204*, 739-740, (1964b).
Brown, P. K. and G. Wald, *Science*, *144*, 45-52, (1964).
Brown, K. T., K. Watanabe and M. Murakami, *Cold Spring Harbor Symposia on Quantitative Biology*, *30*, 457-482, (1965).
Carr, R. E., H. Ripps, I. M. Siegel and R. A. Weale, *Invest. Ophthalmol.*, *5*, 497-507, (1966).
Cohen, A, *J. Cell Biol.*, *37*, 424-444, (1968).
Cone, R. A., *Nature*, *204*, 736-739, (1964).
Cone, R. A., *Cold Spring Harbor Symposia on Quantitative Biology*, *30*, 483-490, (1965).
Cone, R. A., private communication, (1968).
Cone, R. A., *Science*, *155*, 1128-1131, (3 March 1967).
Cone, R. A. and K. T. Brown, *Science*, *156*, 536, (1967).
Crawford, J. M., P. W. Gage and K. T. Brown, *Vision Res.*, *7*, 539-551, (1967).
Ebrey, T. G. and R. A. Cone, *Nature*, *213*, 360-362, (1967).
Enoch, J. M. and W. S. Stiles, *Optica Acta*, *8*, 329-358, (1961).
Fatechand, R., private communication, (1968).
Goldstein, B., *Vision Res.*, *6*, 39-50, (1966).
Hsia, Y. and C. H. Graham, *Proc. Nat. Acad. Sci. U. S.*, *43*, 1011-1019, (1957).
Hubbard, Ruth, *N. P. L. Symposium on "Visual Problems of Colour", H. M. S. O. London*, *1*, 151-169, (1958).
Hubbard, Ruth, D. Bownds and T. Yoshizawa, *Cold Spring Harbor Symposia on Quantitative Biology*, *30*, 301-315, (1965).
Hubbard, Ruth and A. Kropf, *Proc. Nat. Acad. Sci. U. S.*, *44*, 130-139, (1958).
Matthews, R. G., Ruth Hubbard, P. K. Brown and G. Wald, *J. Gen. Physiol.*, *47*, 215-240, (1963/4).
Morton, R. A., *Nature*, *153*, 69-71, (1944).
Ostroy, S. E., F. Erhardt and E. W. Abrahamson, *Biochim. Biophys. Acta*, *112*, 265-277, (1966).
Pak, W. L., *Cold Spring Harbor Symposia on Quantitative Biology*, *30*, 493-499, (1965).
Pak, W. L. and R. J. Boes, *Science*, *155*, 1131-1133, (1967).

Pak, W. L., V. P. Rozzi and T. G. Ebrey, *Nature*, *214*, 109-110, (1967).
Pitt, F. H. G., *Med. Res. Council Spec. Rep. Ser.*, *200*, (1935).
Ripps, H. and R. A. Weale, *Vision Res.*, *3*, 531-543, (1963).
Ripps, H. and R. A. Weale, *J. Physiol.*, *173*, 57-64, (1964).
Ripps, H. and R. A. Weale, *Nature*, *205*, 52-56, (1965a).
Ripps, H. and R. A. Weale, *J. Opt. Soc. Amer.*, *55*, 205-206, (1965b).
Ripps, H. and R. A. Weale, *J. Physiol.*, *196*, 67-69P, (1968).
Rushton, W. A. H., *N. P. L. Symposium on "Visual Problems of Colour"*, *H. M. S. O. London*, *1*, 71-101, (1958).
Rushton, W. A. H., *J. Physiol.*, *168*, 345-359, (1963a).
Rushton, W. A. H., *J. Physiol.*, *168*, 360-373, (1963b).
Rushton, W. A. H., *J. Physiol.*, *168*, 374-388, (1963c).
Rushton, W. A. H., *J. Physiol.*, *176*, 24-37, (1965a).
Rushton, W. A. H., *J. Physiol.*, *176*, 38-45, (1965b).
Rushton, W. A. H., *J. Physiol.*, *176*, 46-55, (1965c).
Rushton, W. A. H., *J. Physiol.*, *176*, 56-72, (1965d).
Rushton, W. A. H. and G. H. Henry, *Vision Res.*, *8*, 617-631, (1968).
Schmidt, W. J., *Kolloidztschr.*, *85*, 137-148, (1938).
Stiles, W. S., private communication, (1939).
Villermet, Gabrielle M. and R. A. Weale, *J. Roy. Microscop. Soc.*, *84*, 565-569, (1965).
Villermet, Gabrielle M. and R. A. Weale, *Proc. Roy. Soc.*, *B164*, 96-105, (1966).
Villermet, Gabrielle M. and R. A. Weale, in press, (1969).
Wald, G., P. K. Brown and I. R. Gibbons, *J. Opt. Soc. Amer.*, *53*, 20-35, (1963).
Wald, G. and P. K. Brown, *Cold Spring Harbor Symposia on Quantitative Biology*, *30*, 345-359, (1965).
Walraven, P. L. and M. A. Bouman, *J. Opt. Soc. Amer.*, *50*, 780-784, (1960).
Weale, R. A., *J. Physiol.*, *132*, 257-266, (1956).
Weale, R. A., *Optica Acta*, *6*, 158-174, (1959).
Weale, R. A., *Science*, *145*, 1205-1206, (1964).

Weale, R. A., *Cold Spring Harbor Symposia on Quantitative Biology*, *30*, 335-342, (1965).
Weale, R. A., *J. Physiol.*, *186*, 175-186, (1966).
Weale, R. A., *Vision Res.*, *7*, 819-927, (1967).
Weale, R. A., *Nature*, *218*, 238-240, (1968).
Williams, T. P., *J. Gen. Physiol.*, *47*, 679-689, (1964).
Willmer, E. N., *J. Physiol.*, *110*, 422-446, (1949).
Wolken, J. J., *J. Opt. Soc. Amer.*, *53*, 1-19, (1963).
Yoshizawa, T. and G. Wald, *Nature*, *197*, 1279-1286, (1963).

Radiation Damage and Repair In Vivo

Philip C. Hanawalt

Department of Biological Sciences
Stanford University

The sun is the principal source of energy for the biosphere; thus, photosynthetic processes have been of central importance for the growth and evolution of living systems on the earth. Light supplies the energy for the synthesis of most of the organic molecules of which all life is composed. Unfortunately, the effects of photons in the ultraviolet end of the spectrum (i.e., below about 300 nm) are more often destructive than useful to the necessary functions of these molecules. It is therefore essential that the evolving life forms have been able to circumvent or repair the damaging effects of ultraviolet photons. We will outline some of the known photoproducts that are produced in living systems by ultraviolet light (abbreviated UV), particularly as these are related to cell survival. We will then consider the various means by which cells respond to radiation damage to enhance recovery. Some of the repair mechanisms apply to the damaging effects of ionizing radiations and various chemical mutagens as well. Finally, we will discuss some indications that the story is far from complete; there may be existent repair mechanisms that are yet to be discovered.

Let us begin by considering in very general terms the possible means by which a living system might respond to the presence of photochemically altered or otherwise damaged molecules. Three possible modes for dealing with such damaged molecules could be designated as follows:

1. The damaged molecule (or portion of a molecule) might be restored to its functional state *in situ*. This could be achieved by the simple "decay" of the damage to an innocuous form, or it might be the result of an enzymatic repair mechanism.
2. The damaged unit could be removed from the molecule or system and then replaced with an undamaged unit to restore the functional state of the system.
3. The damage could remain unrepaired in the system, but the system might be able to bypass or ignore the damage.

All three of these recovery modes have now been documented in various living systems.

Which molecules are essential to the survival of the living system? The information-containing macromolecules, the proteins, and the nucleic acids comprise as much as 90% of the dry weight of some cells. The proteins are polymers made up of amino acids, and a simple bacterial cell may contain several thousand different protein molecules. However, it also may contain a thousand identical proteins of a particular type. The destruction of a number of the proteins of a given type would probably not kill the cell because of the multiplicity of identical copies of that type. The synthesis of proteins requires the translation of the genetic code from sequences of nucleotide subunits in nucleic acid polymers to the corresponding amino acid sequences in the proteins. This is accomplished by specialized nucleic acids and enzymes (i.e., proteins that function as catalysts) for which there are many identical copies in the cell. Thus, some elements of the translation machinery could also be destroyed without totally disabling the cell. However, the source of the genetic specification of a protein is in the primary genome of the cell. This information is encoded in the nucleotide sequences in the desoxyribonucleic acid (DNA) of the chromosomes, and there may be only a few copies of this information. At certain stages of growth a bacterial cell may contain only a single copy of the genome. As long as the information in the genome is intact, the cell will generally be able to replace damaged molecules of other types as needed for continued growth and survival. On the other hand, it is clear that damage to the DNA is more serious and can lead to a permanent hereditary change (i.e., mutation) or even death. One would, then, suspect that the DNA might be the most sensitive target for the action of some deleterious agents on cells.

Of course, not all such agents act directly on DNA. In addition many chemical substances that might attack DNA are prevented from entering the cell by the permeability restrictions of cell membranes. These membranes pose no barrier to photons, and this has facilitated the use of monochromatic light as a very specialized probe for studying the intracellular machinery. Visible light absorbed in the carotenoid components in the membranes of some cells can lead to eventual disruption of the membranes and killing of the cells. The principal wavelengths absorbed by the proteins and nucleic acids are in the UV region of the electromagnetic spectrum. It should be remembered that a photon must be absorbed in order to cause a photochemical reaction and a possible resultant biological effect.

The action spectrum for mutation production and for killing of cells follows very closely the nucleic acid absorption spectrum. The characteristic maximum absorption at 260 nm in DNA is due to the purine and pyrimidine bases (Fig. 1), so one would expect that the biological effects are largely the result of photon absorp-

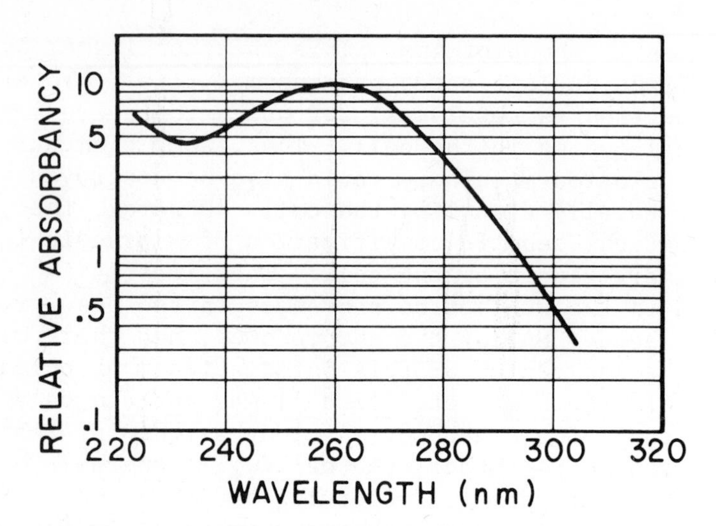

Figure 1. The absorption spectrum of DNA, showing the characteristic maximum at 260 nm. (From "Ultraviolet Radiation and Nucleic Acid" By R. A. Deering. Copyright © 1962 by Scientific American, Inc. All rights reserved.)

tion by the bases. However, the pyrimidines (thymine and cytosine) are more than ten times as sensitive to photochemical alteration as the purines (adenine and guanine).

The current surge of interest in the UV photochemistry of DNA was sparked by the discovery by Beukers and Berends in the late 1950's that thymine in frozen solution forms covalent dimers when irradiated by UV. It was soon found that this photoproduct was also produced in UV-irradiated DNA. In DNA the dimer was formed between adjacent thymines in the same strand. The chemical structure of a cyclobutane dimer of thymine is shown in Fig. 3. Other pyrimidines also form dimers. Thus, cytosine dimers and mixed dimers of thymine and cytosine are found in UV-irradiated DNA. And uracil-containing dimers are induced by UV in ribonucleic acid (RNA). The efficiency of formation of these dimers is dependent upon the wavelength of the irradiation. The action spectrum for their formation follows the absorption spectrum of DNA (Fig. 2). They can, however, be broken by UV to yield the free

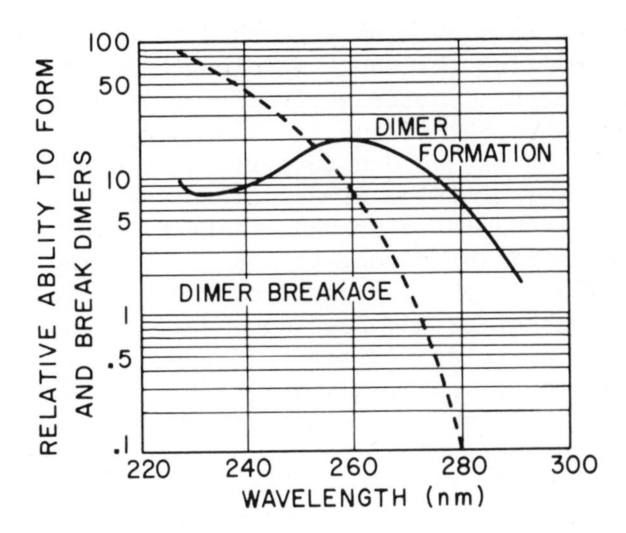

Figure 2. Action spectra for the formation and breaking of thymine dimers. Irradiation of DNA at a given wavelength eventually results in a characteristic equilibrium level of dimerized thymine. This level is lower for 240 nm irradiation than for 260 nm irradiation. (From "Ultraviolet Radiation and Nucleic Acid" By R. A. Deering. Copyright © 1962 by Scientific American, Inc. All rights reserved.)

thymines again. At a given wavelength an equilibrium is eventually attained for the formation and splitting of dimers. Although this equilibrium level is lower for 240 nm irradiation than for 260 nm irradiation, it still represents an amount of dimerized thymine that is too high to be tolerated by a viable cell. The effect of dimers in DNA is probably analogous to the effect of fusing together two adjacent teeth of a zipper. The normal replication of DNA requires that the two complementary parental strands unwind as daughter strains are synthesized. The dimerization of adjacent bases places them too close together for proper fitting into the double-stranded DNA structure. If a transforing DNA is irradiated at 260 nm, it loses biological activity. The reirradiation of this DNA at 240 nm restores some of the activity, indicating that pyrimidine dimers must be involved in the inactivation process.

The Setlows have been able to show by an elegant approach combining the short wavelength reversal of dimers and enzymic photoreactivation (discussed below) that more than 50% of the inactivation of a transforming DNA by UV may be due to pyrimidine dimers.

However, pyrimidine dimers are not the entire story in the photochemistry of DNA. Figure 3 shows some of the other sorts of damage to DNA in UV-irradiated cells. The hydration product of cytosine was discovered by Sinsheimer and Hastings nearly twenty years before the pyrimidine dimer made its debut. In an *in vitro* system it has been shown that the hydration products of pyrimidines can simulate a mutation effect. That is, the hydration product of cytosine behaves as though it were uracil as far as the reading of the genetic material in transcription is concerned. The hydration products are not generally formed in great yield in double-stranded DNA, however, and they are also quite unstable, reverting spontaneously to the original pyrimidine form. Both the dimers and the hydration products would affect the hydrogen bonding properties of the pyrimidines in the DNA and would result in local regions of denaturation. Still another sort of photochemical alteration in DNA entails the covalent cross-linking of DNA to protein. Little is known about the biological significance of this kind of photochemical product, although it has been implicated in the UV inactivation of some systems that are extraordinarily resistant to the production or effects of thymine dimers. The backbone break illustrated in Figure 3 is one of the likely consequences of ionizing radiation, and the break also may occur indirectly as a result of the repair processes that follow the production of pyrimidine dimers in DNA, as discussed below.

The relative importance of different photoproducts to biological inactivation is strongly dependent upon the physical and physiological state of the system. In dry bacterial spores, for example, no cyclobutane-type thymine dimer is found in the DNA by UV, but instead, a new thymine photoproduct is found. At very low temperatures less pyrimidine dimers are formed than at 37°C and the action spectrum for bacterial in-

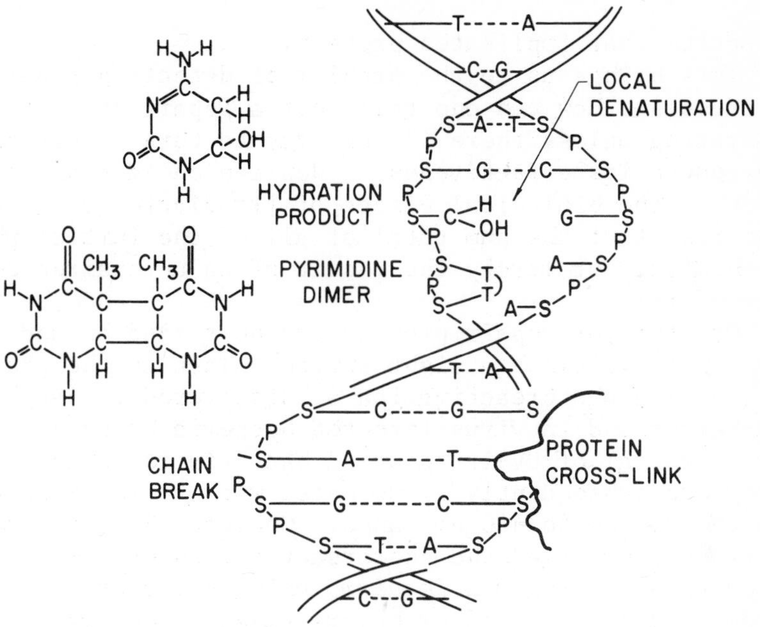

Figure 3. Schematic illustration of the various alterations found in DNA extracted from cells that have been irradiated with ultraviolet light. Some of these physical changes (e.g., denaturation) may be secondary effects resulting from photoproducts such as thymine dimers. The dimerized thymines are evidently unable to maintain hydrogen bonds with the adenine in the complementary strand. Other effects such as cross-linking to protein may involve new photoproducts yet to be characterized. Chemical structures for the cytosine hydrate and the thymine dimer are shown. (From "Ultraviolet Radiation and Nucleic Acid" By R. A. Deering. Copyright © 1962 by Scientific American, Inc. All rights reserved.)

activation begins to implicate protein as well as nucleic acid. Also, the bacterium *Micrococcus radiodurans*, which has a very efficient repair mechanism, is inactivated after very high doses of UV with an action spectrum that implicates protein.

This brings us to the problem of detecting repair processes. How can one tell that a repair process is operating unless there is some way to turn it off or to reduce its effectiveness? How can one determine whether the biological effect of irradiation represents the result of the sum total of all of the initial photochemistry or merely the result of an unrepaired sector of it?

One type of repair process can be turned on and off by simply flicking a light switch. That is the process of enzymic photoreactivation as discovered in bacteria by Kelner and in virus-infected bacteria by Dulbecco. The survival of UV-irradiatied bacteria is enhanced if they are subsequently illuminated with visible light in the 310 nm to 440 nm range. Goodgal, Rupert, and Herriott performed the photoreactivation of UV-irradiated transforming DNA *in vitro* using cellular extracts from *Escherichia coli* or bakers yeast. The enzyme responsible for the repair has now been purified manyfold by Muhammad, but the chromophore is still unknown. The enzyme binds to UV-irradiated DNA in the dark. Upon visible illumination of the DNA-enzyme complex, the pyrimidine dimers are severed *in situ* and the enzyme is released. The visible light actually supplies

---

Figure 4. (Opposite page.) Ultraviolet inactivation and photoreactivation (PR) of the transforming ability of DNA from *Hemophilus influenzae*. DNA isolated from a strain of a given genetic type A can be incorporated into cells of another genetic type B and, by genetic recombination, can confer some of the characteristics of the type A strain upon the type B strain. Ultraviolet irradiation destroys this property of transformation as indicated in the above curve. However, if the irradiated DNA is incubated with an extract from yeast in the presence of visible light (310 to 440 nm), then some of the transforming ability will be recovered. Photoreactivation reduces the effective dose of the ultraviolet irradiation by the enzymic mechanism of splitting the pyrimidine dimers *in situ*. (Adapted from J. K. Setlow, *Rad. Res. Suppl.*, *6*, 141, 1966.)

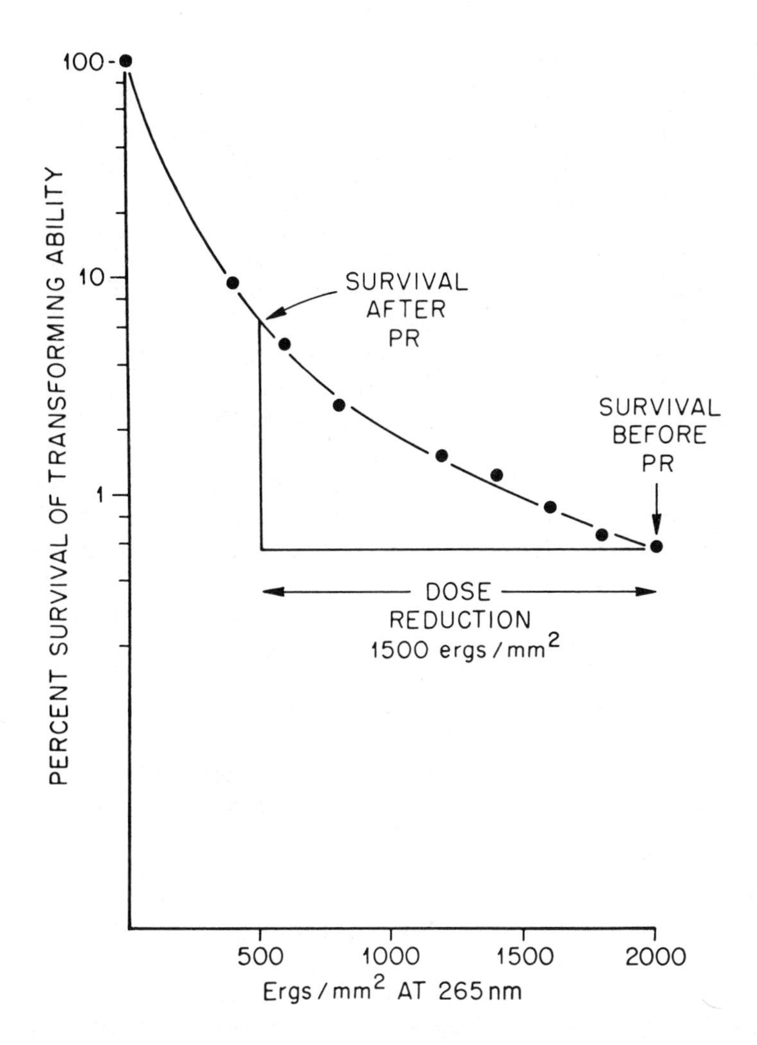

the energy for the dimer splitting. (It is not just that the enzyme needs light to see what it is doing!) No photoproducts other than pyrimidine dimers in DNA are known to be repaired by the photoreactivating enzyme system.

Figure 4 illustrates the effect of enzymic photoreactivation on the survival of biological activity in transforming DNA. There are some definite advantages to studying such processes with transformation as the biological assay. The isolated DNA can be irradiated in the absence of any enzymes or factors that might alter the initial photochemical effect. The chemical nature of the damage can be assessed. The DNA can be exposed to other factors such as cell extracts, under controlled conditions. Then, the DNA can be placed into a viable biological system (e.g., bacterium) for a very sensitive indication of the biologial effect. The effect of exposure of the irradiated DNA to photoreactivating enzyme and visible light is that of an apparent reduction in the UV dose.

The process of enzymic photoreactivation is surprisingly widespread. The simplest living cells, the mycoplasma (with only one fifth as much DNA in their genomes as *Escherichia coli*), can perform photoreactivation. A variety of more complicated cellular systems also contain photoreactivating activity. Some of these are: the protozoan *Tetrahymena pyriformis;* the unicellular green alga, *Euglena gracilis;* frogs, reptiles, and some fish; chick embryos; the sea urchin; the Australian kangaroo and other marsupials; etc. However, photoreactivation has not been demonstrated in mammalian systems or in tissue cultures. Nor is there certain evidence that any RNA system, such as Tobacco Mosaic Virus, is subject to enzymic photoreactivation although viability can sometimes be enhanced by visible illumination of the leaf after infection with the UV-irradiated virus. This latter effect of visible light may be an indirect one rather than a repair of photochemical lesions. A process known as indirect photoreactivation has been shown to operate in bacterial mutants that are deficient in enzymic photoreactivation, and it presumably functions by enhancing the efficiency of a dark repair process.

It was more difficult to pin down the molecular nature of the repair processess that operate in the dark. Yet, there were early indications that such repair processes must exist. In fact, the first indication of possible recovery phenomena in connection with photochemical damage came from the studies of Hollaender and Claus in 1936, when they found that higher survival levels of UV-irradiated fungal spores could be obtained if they were placed in liquid media for a period before being allowed to grow on a nutrient agar surface. More than ten years later, Roberts and Aldous extended these observations by showing that the shapes of the UV survival curves for some bacterial strains could be changed quite drastically by varying the culture growth conditions *after* the irradiation.

Another strong line of evidence for the existence of dark repair processes was the discovery of radiation-sensitive mutants of bacteria. The first one, *Escherichia coli* strain B_{s-1}, was isolated by Ruth Hill. Soon others were reported in K-12 strains of *E. coli*. These mutants were found to yield reduced levels of virus upon infection with UV-irradiated virus than the original parent strains. Thus, it was postulated that some repair function in the bacterium normally performed repair of damaged virus DNA. (This process was termed host cell reactivation.) Genetic mapping experiments conducted by Adler and by Howard-Flanders showed that certain loci in the genome were responsible for the radiation sensitivity phenomena. Yet, no detectable difference in the photochemical effects of UV upon different bacterial strains could be ascertained; that is, the same dose of UV produced about as many thymine dimers in *E. coli* B_{s-1} as in the parent radiation resistant strain.

It was found that some of the physiological effects that enhanced the survival to irradiation were more pronounced in the resistant strains than in the sensitive mutants. Thus, it was conjectured that such effects (like holding cells in buffer after irradiation before placing them in nutrient medium) might function by inhibiting normal growth processes while the repair systems completed their task. It was intuitively evident that the cell should have a better chance of survival if the damage in its DNA were repaired before the damaged regions attempted the normal functions of replication and transcription. In support of this idea was the finding that cells that had completed their normal DNA replication cycles were strikingly more resistant to UV than cells that were in the midst of the cycle at the time of irradiation. (Inhibition of protein synthesis in bacteria allows cells to complete their DNA replication cycles, but not to start new ones.)

This physiological effect is shown in Fig. 5(a). The simple exponential survival curve for the growing culture has been interpreted, on the basis of target theory, to mean that the inactivation of certain sensitive sites (i.e., in the DNA) leads to cell death. Survival curves with shoulders have been interpreted to mean that there are a number of sensitive sites and that *each* site must be inactivated to cause the death of the cell. The kinetics for the so-called multitarget model are indicated by the dashed line in Fig. 5(a). The actual data are more readily interpreted to mean that at low doses a large fraction of the damage can be repaired (if the cell is not replicating its DNA). At higher doses the repair is less effective, either

---

Figure 5. (Opposite page.) The effect of physiological state upon sensitivity to killing by ultraviolet light of two bacterial strains. (a) Strain *E. coli* TAU is able to recover from photochemical damage in the dark. A rapidly growing culture exhibits a simple exponential survival curve as shown (●). The cells appear strikingly more resistant if growth has been inhibited by withholding essential amino acids (▲). It is thought that the dark repair processes may be much more efficient if normal growth is inhibited until repair is completed. The dashed line indicates the curve shape that would be expected if the enhanced resistance were alternatively due to the presence of a multiplicity of sensitive targets in each cell such that all must be hit by photons to cause inactivation of the cell. (b) The strain *E. coli* B_{s-1} has been shown to be deficient in the dark repair of damaged DNA. A growing culture is much more sensitive to photochemical inactivation than the strain *E. coli* B from which it was derived. The inhibition of protein synthesis in this strain yields an ultraviolet survival curve (▲) that differs only slightly from that exhibited by a rapidly growing culture (●). Presumably this strain does not benefit from the inhibition of normal growth because the repair process is nonfunctional. (Adapted from P. C. Hanawalt, *Photochem. Photobiol.*, 5, 1, 1966.)

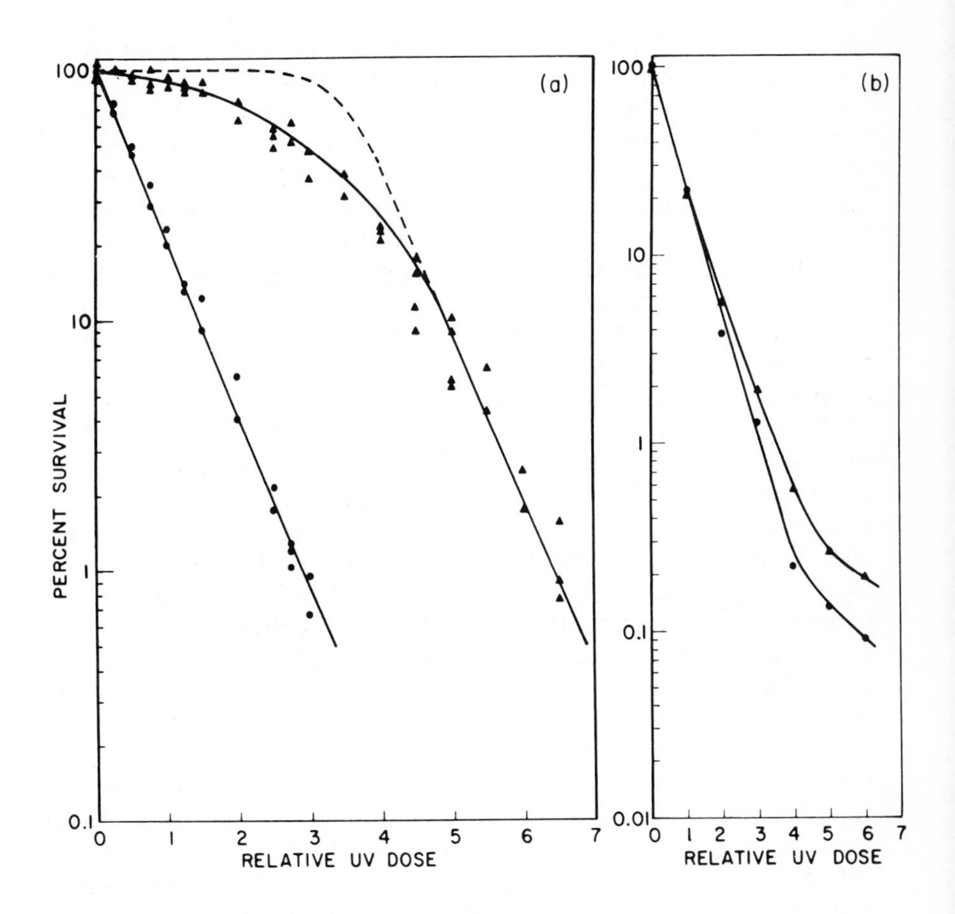

(a)
100
10
1
0.1
PERCENT SURVIVAL
0
1
2
3
4
5
6
7
RELATIVE UV DOSE
(b)
100
10
1
0.1
0.01
0
1
2
3
4
5
6
7
RELATIVE UV DOSE

because the repair system becomes saturated or possibly because the radiation is inactivating the repair system itself. In contrast to the result in Fig. 5(a), note in Fig. 5(b) that the inhibition of protein synthesis in the UV-sensitive strain B_{s-1} does not have an appreciable effect upon survival. Evidently the inhibition of DNA synthesis provides no advantage for a cell that is deficient in the dark repair process.

What is the mechanism of this dark repair process? Is it also an enzyme that splits pyrimidine dimers? The answer is "no" to the latter question. Rather, the repair mechanism involves the utilization of an intact strand of DNA to recover the information in the complementary damaged one. The base pairing relationships in the complementary strands are quite specific. Thus, a guanine (G) in one strand must always be opposed by a cytosine (C) in the other; and likewise, an adenine (A) in one strand must be hydrogen bonded to a thymine (T) in the complementary strand (See Fig. 3). If one were to damage one of the strands so that the information were lost or no longer recognizable, then in principle that information could be reconstructed from the complementary information in the other strand. That is precisely what happens in the dark repair scheme, and the existence of such a mechanism provides a possible explanation for the evolution of a double-stranded (redundant) form for the genetic blueprint in the cell.

The first indication of the molecular mechanism for this repair process came from the studies of R. Setlow and Carrier, who studied DNA synthesis in UV-irradiated resistant bacteria and in a UV-sensitive mutant. They found that the resistant strains eventually recovered from the UV inhibition of DNA synthesis while the sensitive strain did not. The thymine in the DNA was radioactively labeled with tritium, and the fate of the UV-induced thymine dimers was determined. (It is possible to distinguish free thymine from the dimerized form by the technique of paper chromatography.) They found that thymine dimers were released from the DNA in the resistant cells but not in the sensitive cells. Similar observations were reported for other combinations of UV-resistant and UV-sensitive strains of bacteria by Boyce and Howard-Flanders and by Riklis. These observations led to a model for the repair process that has come to be known colloquially as "cut and patch" and which involves the excision of the damaged DNA segment followed by reconstruction of that segment. The postulated steps in this process are listed in Fig. 6. Direct physical evidence for the repair replication step was obtained by Pettijohn and Hanawalt by the density-labeling method outlined in Fig. 7. Some steps in the process have now been demonstrated *in vitro*. Cellular extracts from *Micrococcus lysodeikticus* have been shown to perform the selective excision of pyrimidine dimers from UV-irradiated DNA. Also, an enzyme known as polynucleotide ligase has been purified, and it possesses the necessary specificity for the final rejoining step in which the repaired segment must be covalently joined to the contiguous parental DNA strand. The DNA polymerase isolated by Kornberg might well be the one responsible for the repair synthesis step in *E. coli*, and this enzyme has been shown to perform a repair replication function in a model system *in vitro*.

As yet the actual sequence of steps is not certain. The repair replication step might begin as soon as the incision is made adjacent to the damage, as indicated in the alternative step III' (Fig. 6). The damaged region could then be peeled away and eventually detached from the parental strand. Repair replication is not observed in the UV-sensitive *E. coli* B_{s-1}, which can not excise dimers. It is also not seen following photoreactivation, as expected if the dimers have already been split *in situ*. The subsequent normal replication of DNA that has undergone repair replication has been demonstrated. Finally, a number of bacterial mutants have been found that are deficient in normal DNA replication at 42°C, but which can still perform repair replication at that temperature. This suggests that the normal and the repair DNA polymerase systems are not identical.

Evidence for the excision-repair scheme has now been

Figure 6. (Opposite page.) Schematic representation of the postulated steps in the excision-repair of damaged DNA. Steps I through VI illustrate the "cut and patch" sequence. An initial incision in the strand containing the damage is followed by local degradation before the resynthesis of the deleted region has begun. In the alternative "patch and cut" model the resynthesis step III' begins immediately after the incision step II, and the excision of the damaged region occurs when repair replication is complete. In either model the final step VI involves a rejoining of the repaired section to the contiguous DNA of the original parental strand.

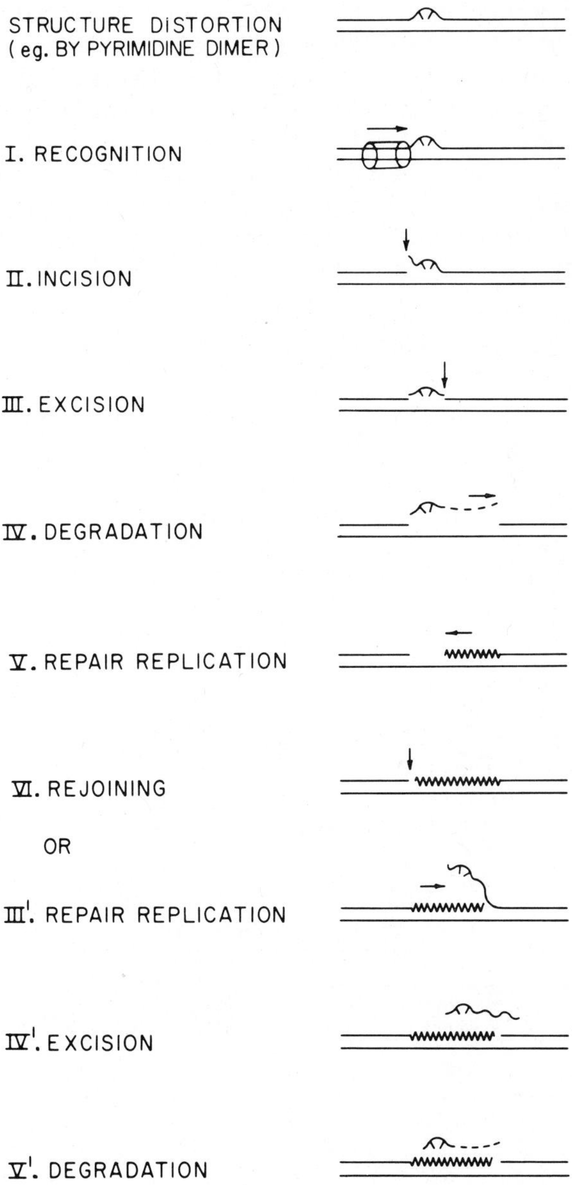

STRUCTURE DISTORTION
(eg. BY PYRIMIDINE DIMER)

I. RECOGNITION

II. INCISION

III. EXCISION

IV. DEGRADATION

V. REPAIR REPLICATION

VI. REJOINING

OR

III'. REPAIR REPLICATION

IV'. EXCISION

V'. DEGRADATION

obtained in a number of different cell types. The fact that the mycoplasmas perform repair replication attests to the importance of this mode of repair in even the simplest of cells. (In fact, the larger bacterial viruses also contain information in their genomes for the synthesis of repair enzymes.) The protozoan *Tetrahymena pyriformis* performs repair replication after UV irradiation. The radioresistant organism *Micrococcus radiodurans* has an extremely efficient excision-repair system, a fact which may explain why it does

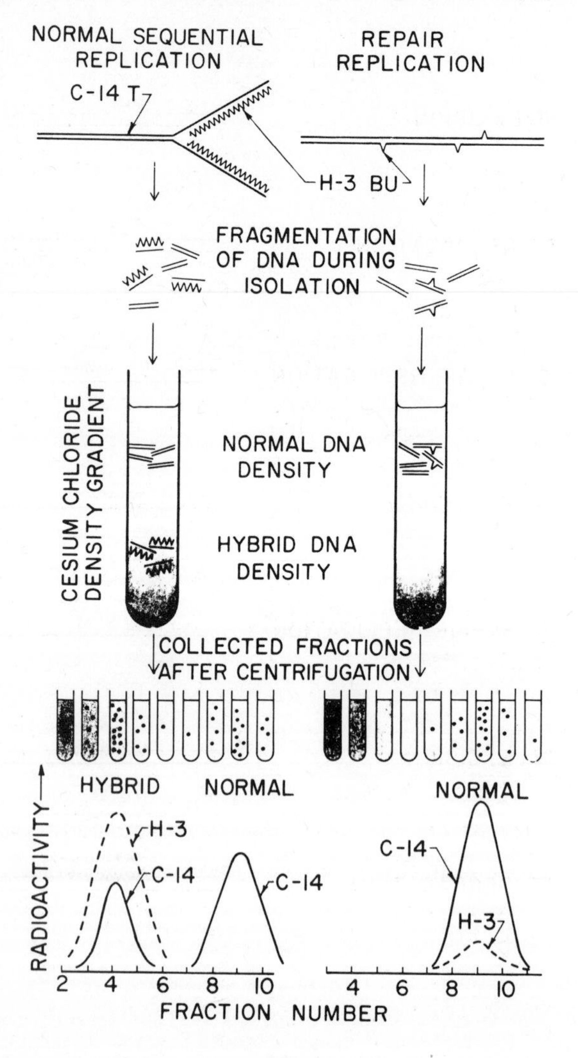

not have photoreactivating enzyme activity. Most exciting has been the recent examination of mammalian cells in tissue culture. Regan and Trosko demonstrated the preferential removal of thymine dimers from the DNA of three human cell lines in culture, and Rasmussen and Painter have used the 5-bromouracil labeling method to indicate repair replication in HeLa cells. Cleaver has

---

Figure 7. (Opposite page.) Protocol for the demonstration of normal replication and repair replication of DNA in growing cells. The DNA is first radioactively labeled (e.g., by the incorporation of carbon-14 labeled thymine); then the cells are permitted to incorporate another radioactive label at the same time that a "density label" is being incorporated (e.g., hydrogen-3 labeled 5-bromouracil). 5-bromouracil (BU) is an analogue of thymine that can be incorporated into DNA in place of the natural base thymine. It has the effect of increasing the density of the DNA fragments that contain it. This density increase is, of course, proportional to the relative amount of thymine and BU in the DNA. Thus, parental DNA fragments that contain short regions of repair may differ little in density from those that contain no BU, while normally replicated (hybrid) fragments that contain 50% BU will be much denser. The density distribution of the isolated DNA fragments is analyzed by means of equilibrium sedimentation in the ultracentrifuge in a density gradient of cesium chloride. At equilibrium the DNA fragments will be found in the gradient at positions that correspond to their buoyant densities rather than to their size. This is essentially the method developed by Meselson, Stahl, and Vinograd and utilized to prove that DNA normally replicates semiconservatively (shown on left half of figure). The collected fractions are examined for the presence of parental label (C-14) as well as the density label (H-3). Parental DNA fragments that contain short regions of repair may differ little in density from those that contain no 5-BU (Shown on right half of figure). (From "The Repair of DNA" by P. C. Hanawalt and R. H. Haynes.)

applied the density labeling method to study repair replication in skin fibroblasts from normal humans and from patients with a rare hereditary skin disease known as *Xeroderma pigmentosum*. This disease results in an extreme sensitivity to the induction of skin cancers by sunlight. Cleaver found that normal skin fibroblasts could perform repair replication but that those from diseased patients could not. This is perhaps the first suggestion that the excision-repair process may be important in the resistance of cells to cancerous transformation.

Unlike photoreactivation, the excision-repair process is capable of recognizing the repairing of a variety of structural defects in DNA. For example, repair replication is observed following treatment of bacteria with a compound related to mustard gas, nitrogen mustard, which attacks guanine rather than thymine in DNA. Repair is also seen after treatment of cells with other alkylating agents such as methyl methane sulfonate or the powerful mutagen, nitrosoguanidine. The recognition of damage probably involves the detection of some gross distortion of the backbone structure of the DNA rather than the identification of specific base defects. The evidence for the excision-repair of damage produced in DNA by ionizing radiations is less certain, possibly because of the generally greater amount of degradation of DNA resulting from these radiations.

We have mentioned above that X-rays produce breaks in DNA. The best evidence that these breaks can be repaired *in vivo* is seen in the method of McGrath and Williams illustrated in Fig. 8. The size of isolated single-stranded fragments of DNA from irradiated bacteria is smaller than those from nonirradiated cells. In the course of continued incubation, however, it can be seen that the fragments increase in size to the distribution expected from unirradiated cells. This same approach can be used to follow the incision and eventual rejoining steps in the excision-repair scheme after UV irradiation of bacteria. Some bacterial mutants have been shown to be sensitive to UV, but not to X-rays. This could be understood in terms of the excision-repair scheme if it were assumed that some mutants lacked only the first recognition step in the process and that the other steps could proceed following the introduction of single-strand breaks by X-irradiation.

One might wonder how X-rays kill cells if the single-strand breaks in DNA can be repaired efficiently? The answer, in part, is that there are other types of damage produced, as summarized in Fig. 9. Some of these other types are not repairable. Although some of the action of ionizing radiation involves direct interaction with the DNA, most of the energy is normally dissipated by the production of highly reactive free radicals in the aqueous environment surrounding the DNA. These free radicals and resultant peroxides can then attack the DNA and produce a variety (as yet uncatalogued) of damage to the bases as well as strand breaks. Although single-strand breaks and at least some of the base damage may be repairable, there is no known mechanism that can repair a double-strand break in DNA. (In fact, it would be hard to imagine what sort of mechanism could conceivably join a broken double-strand end to the *appropriate* double-strand end from which it had been severed. The complementary base-pairing relationships would be of no value in such a case.) The killing of a bacterial virus by X-rays has been shown by Freifelder to correlate directly with the introduction of double-strand breaks in the DNA. One double-strand break is sufficient to kill the virus. Kaplan has also shown by the method of McGrath and Williams that the killing of bacteria by X-rays correlates roughly with the production of double-strand breaks rather than with the production of single-strand breaks, and that the double-strand breaks are not repaired.

At this point we have documented two of the three modes for dealing with damage: photoreactivation restores dimerized pyrimidines to their normal state *in situ*, and excision-repair involves the replacement of damaged nucleotides with undamaged nucleotides to restore normal function to the DNA. How about the third mode? Can the cell bypass or ignore damage in its DNA?

The possibility of ignoring the damage is exemplified

in the case of a polyploid cell. A polyploid cell has more than one copy of the genome, and it can often survive when one copy is rendered nonfunctional. It is also possible for two damaged genomes to cooperate to restore function to the cell. The process of genetic recombination involves the physical exchange of segments of DNA between genomes (although this exchange is generally unilateral). Two UV-inactivated viruses may undergo recombination in the host cell so as to produce one virus that is viable. This process is called multiplicity reactivation, and it is a fortuitous sort of recovery process that does not require any recognition

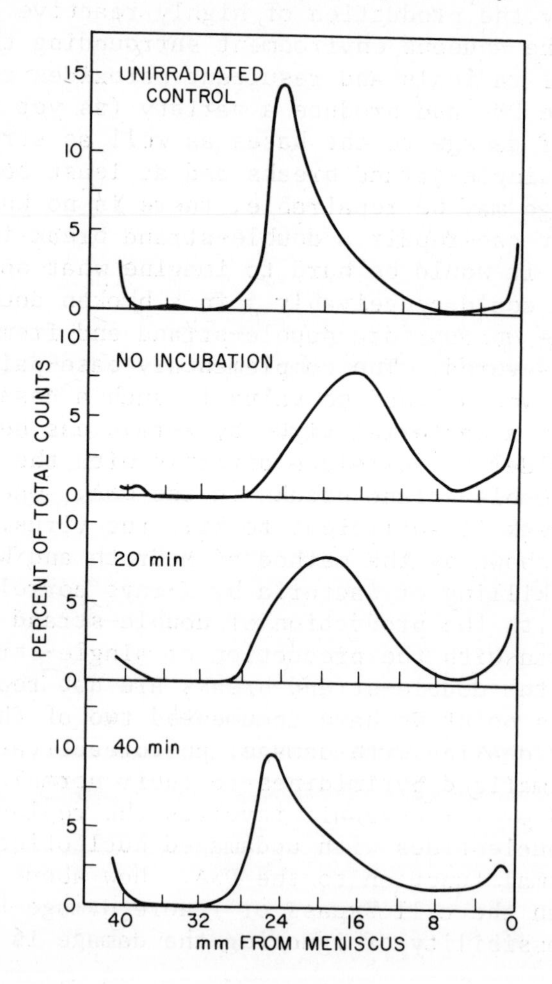

of the damage (although UV irradiation has been shown to increase the frequency of genetic recombination). A "good" segment of DNA from one damaged virus may fortuitously exchange with a "bad" segment from the other virus to produce one intact functional virus.

Genetic recombination may also play a role in the recovery of uninfected bacteria from UV damage. Several years ago a new class of bacterial mutants was discovered by Clark and Margulies. These mutants were found to be deficient in genetic recombination, and they were also found to be UV-sensitive. It was at first supposed that some step in the excision-repair scheme was deficient, some step that was perhaps common to both repair and to genetic recombination. It now appears, however, that the recombination function is concerned with still another dark repair system that may be equally as interesting as the excision-repair one. Recombination-deficient (rec-) cells have

---

Figure 8. (Opposite page.) Demonstration of the repair of single-strand breaks in DNA following irradiation of bacterial cultures with X-rays. A culture of *E. coli* strain B/r was irradiated with a 20 kr dose. DNA was gently isolated from cells immediately following irradiation or after allowing different periods of growth following irradiation. The size distribution of the single-strand fragments of DNA was analyzed by sedimentation in the ultracentrifuge in an alkaline (to intentionally denature the DNA) sucrose gradient. Longer fragments sediment more rapidly and are found farther from the meniscus when fractions are collected at the end of the run as shown above. Without incubation the DNA fragments from irradiated cells are significantly shorter than those from an unirradiated control culture. After 40 min of incubation, however, it is evident that the DNA fragments from irradiated cells have regained the same size distribution obtained from the control culture. The curves shown were fitted to combined data from three experiments, although the points are not shown in this drawing. (Adapted from R. A. McGrath and R. W. Williams, *Nature, 212,* 534, 1966.)

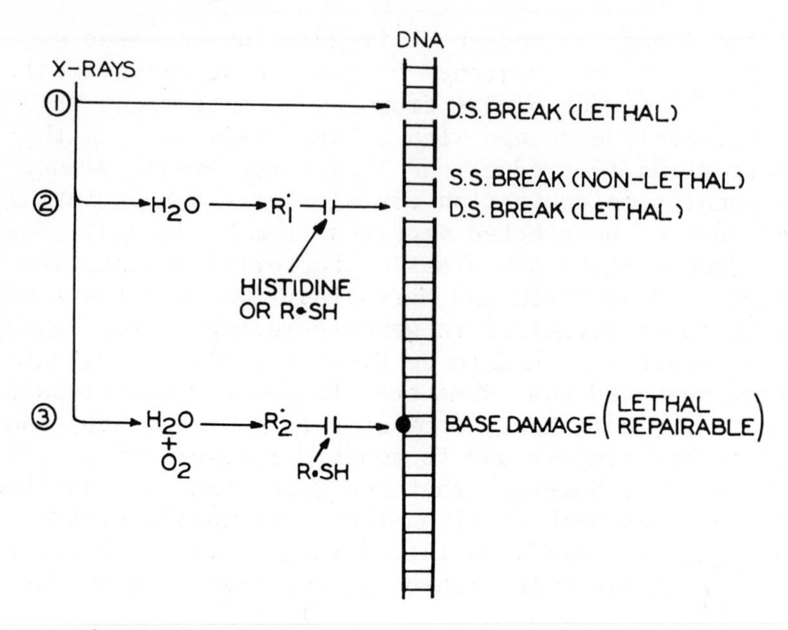

Figure 9. Diagrammatic representation of the sorts of damage produced by the action of ionizing radiation on DNA. (1) Direct production of single or double-strand breaks. (2) Indirect production of breaks as mediated by radiation-induced free radicals. Radical scavengers such as histidine or compounds containing sulfhydryl groups may quench this action. (3) Another indirect action dependent upon the production of free radicals that interact with the pyrimidine bases. Double-strand breaks are probably always lethal events, whereas single-strand breaks and some base damage may be subject to dark repair processes. (Adapted from W. Szybalski, *Rad. Res. Suppl.*, *6*, 95, 1966.)

been shown to perform both dimer excision and repair replication steps in excision-repair. The best indication that the rec- function is a new repair system is seen in a comparison of the UV survival curves for mutants deficient in excision-repair and double-mutants that are deficient in both excision-repair and in recombination. It can be seen in Fig. 10 that the

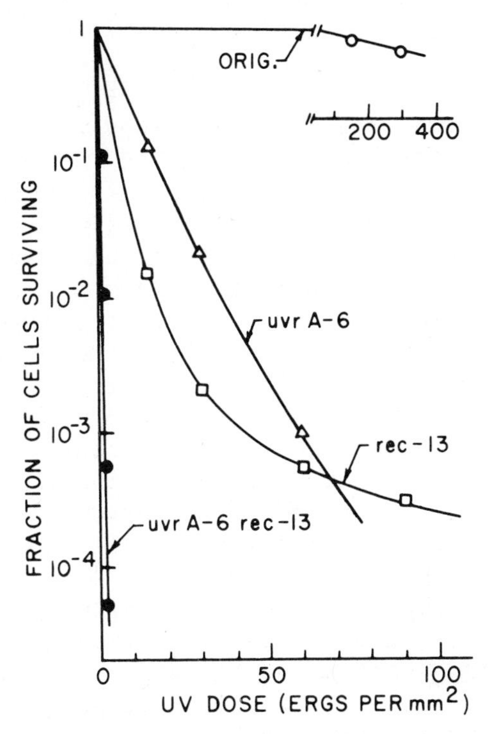

Figure 10. The sensitivity of colony forming ability to ultraviolet light in several UV-sensitive bacterial mutants. The mutant strain uvr A-6 is unable to excise thymine dimers, and it is consequently much more sensitive than the original strain from which it was derived. The mutant rec-13 is defective in the ability to form genetic recombinants, and it is more sensitive than the original strain. The double-mutant uvr A-6 rec-13 is deficient in both excision and recombination, and it is more sensitive than either single mutant. (Adapted from P. Howard-Flanders and R. P. Boyce, *Rad. Res. Suppl., 6,* 156, 1966.)

double-mutant is more sensitive than either single mutant.

A clue to the mechanism of the new repair scheme was found by Rupp and Howard-Flanders in their studies on the nature of the small amount of DNA that is synthesized after UV in bacteria that are deficient in excision-repair. Using the alkaline sucrose gradient method of McGrath and Williams, they were able to show that DNA fragments were synthesized in UV-irradiated cells in lengths roughly corresponding to the predicted distances between pyrimidine dimers in the parental strands. Upon further incubation these newly-synthesized strands were incorporated into large-size pieces of DNA. It has been postulated that another repair system operates in the absence of dimer excision. This process may entail the slow synthesis of DNA around the dimers (a bypass mechanism), or it may somehow utilize genetic recombination to restore integrity to the genome. The elucidation of this mechanism stands as one of the most exciting of the current areas of research on the response of living cells to damage in their DNA.

So far we have been discussing principally the enhancement of survival by the operation of repair mechanisms on DNA. What about the effect of repair processes in relation to the mutagenic effects of UV? One might wonder whether mistakes are ever made in base pairing in the course of repair replication, so that mutations could be the result of faulty repair. It is not likely that this is so, but there are a number of puzzling and perplexing observations. There is good evidence that at least some of the mutagenic effect of UV is due either directly or indirectly to the production of pyrimidine dimers (e.g., photoreactivation reduces mutant yield; mutants deficient in the excision-repair process are more sensitive to mutagenesis as well as to killing, etc.). Thus, one would suppose that at least some of the mutagenic damage produced in DNA by UV can be repaired. There exists, however, a particular class of bacterial mutant that is UV-sensitive, but that does not yield any mutants at all of any type when irradiated with UV. We seem to be faced with a paradox in which mutagenic damage can be perfectly repaired but lethal damage cannot. This is perhaps a good place to stop, but with the prediction that the detailed study of UV mutagenesis may result in the discovery of yet additional mechanisms for dealing with photochemical damage in living systems.

REFERENCES

Hanawalt, P. C. and R. H. Haynes, The repair of DNA. *Sci. Amer.*, *216*, 36 (1967). (Offprint #1061, W. H. Freeman and Company, San Francisco.)

Haynes, R. H., S. Wolff and J. Till, eds., Conference on structural defects in DNA and their repair in microorganisms. *Rad. Res. Suppl.*, *6* (1966).

Howard-Flanders, P., DNA repair. *Ann. Rev. of Biochem.*, *37*, 175 (1968).

Smith, K. C. and P. C. Hanawalt, *Molecular Photobiology: Inactivation and Recovery*, New York: Academic Press (1969).

Witkin, E. M., Mutation-proof and mutation-prone modes of survival in derivatives of *Escherichia coli* B differing in sensitivity to ultraviolet light. In *Recovery and Repair Mechanisms in Radiobiology*, Brookhaven Symposia in Biology #20, p. 17 (1967).

Effects of Ultraviolet Light on the Skin

A. Wiskemann

University Clinic in Dermatology, Hamburg

The effects of ultraviolet light on the skin will vary with the energy distribution of the spectra. Although radiation from the sun and sky is of great personal interest to us, it is less suitable for experimental purposes; and more often the mercury vapor lamp is used with its line spectra filtered or unfiltered. To determine the biological effectiveness of different wavelengths of light, the most useful apparatus is the high pressure xenon lamp used in connection with the monochromator.

The structures which are irradiated in the skin surface can be identified in stained sections of the skin and can be divided into three layers from the surface downwards (Fig. 1).

1. The stratum corneum consisting of dead horny cells without nuclei, from 20 to 80 μ in total thickness. The lowermost layer acts as a barrier

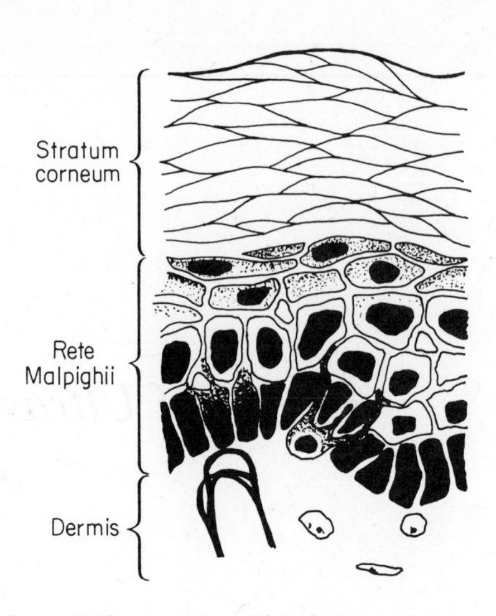

Figure 1. Schematic drawing of the three cellular layers in human skin.

to the diffusion of water in either direction.

2. The Malpighian layer consists of living keratinocytes and is about 25-30 μ in thickness. From its basal layer the skin becomes renewed by cell division and differentiation. Melanocytes, which produce melanin and inject melanin granules into the keratinocytes through their dendrites, are present among the basal cells. The differentiating keratinocytes, with their contained melanin granules, migrate toward the surface where they break down in the stratum granulosum. The horny layer is built up of these cell relics.
3. The dermis, about 2 mm thick, consists of a network of collagenous and elastic fibres which are lodged in an amorphous ground substance. The dermis is pervaded by blood vessels, whose terminal capillaries feed the epidermal cells. Although this layer contains relatively few cells, the photobiologist is usually most interested in the mast cells, which contain vasoactive substances and proteolytic enzymes. Protrusions of the dermis push up into the epidermis from below like

fingertips. These protrusions can be clearly seen, since they contain capillary loops which in the illustration are filled with india ink.

If one looks at a scheme expressing the penetration of ultraviolet radiation into a fair skin, in which the stratum corneum and Malpighian layers are each 25 μ thick, one must appreciate that these figures ignore the irregular border which exists between the epidermis and dermis (Fig. 2).

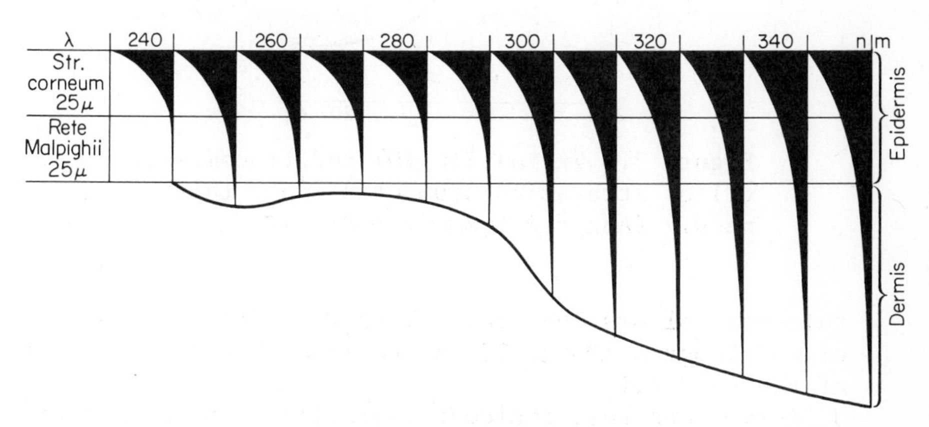

Figure 2. Penetration of the skin by ultraviolet radiation inside a forearm. (Adapted from Tronnier.)

The absorptive power of the different skin layers accounts for the degree of penetration of ultraviolet light into the skin (Fig. 3). Penetration is also governed by scattering, reflection from the surface, and reflection from the deeper layers of the skin. Penetration also varies with the angle of incidence of light and is dependent on such factors as skin color, water content, thickness of the horny layer, etc. Horny layer thickness varies from one region of the body to another and increases after radiation. Suppose the stratum corneum is thickened from 50 to 75 μ. Its transmission is lowered, and, consequently, its extinction is considerably raised. A horny layer 75 μ in

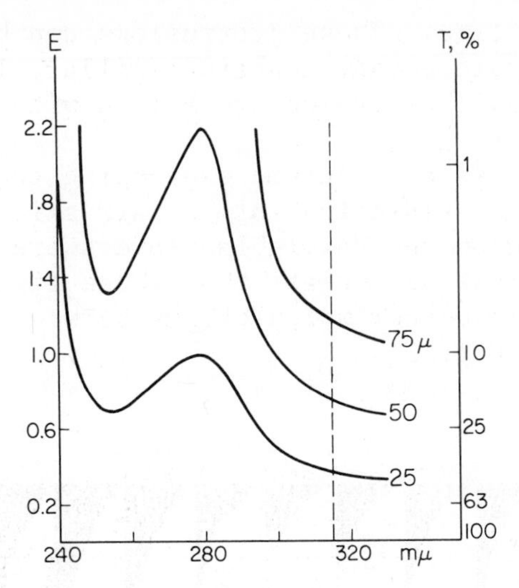

Figure 3. Extinction (E) and transmission (T) of stratum corneum of 25-75 μ thickness. (Adapted from Tronnier.)

thickness allows less than 1% of the incident energy of wavelengths around 254 nm to reach the living cells of the epidermis.

As you can see, incident radiation is mainly absorbed in the stratum corneum. Half of this skin layer consists of scleroproteins and a large quantity of free amino acids. Scleroproteins absorb maximally at 276-278 nm, and this corresponds to the maximum extinction of the horny layer. The maximum absorption of free amino acids eluted from the upper part of the horny layer lies between 265 and 270 nm.

If the upper part of the horny layer is rubbed off with the aid of a strong alkaline solution, the filtrate of this solution, like the horny layer itself, absorbs maximally at 280 nm. If this horny layer is previously irradiated *in vivo* with an ultraviolet spectrum similar to that of the sun, it is found that the peak of extinction curve will be lowered one or two days later. In this manner the horny layer acts on the one hand as a sun screen for the living cells beneath, and on the other hand it gives rise to photoproducts through photochemical reactions within the horny layer, which diffuse downward to act on deeper layers.

In the Malpighian layer absorption will take place through proteins, a type of spheric globulins with a maximum between 275 and 280 nm, and nucleic acids. The absorption spectrum of nucleoproteins is given by the combination of nucleic acid (N) and protein (E) (Fig. 4).

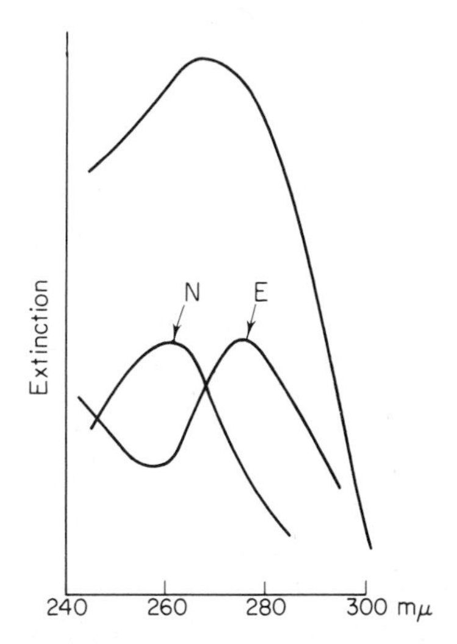

Figure 4. Absorption spectra of nucleic acid (N), protein (E), and nucleoprotein (upper curve).

We know very little of the sites in the cells and the tissues at which primary photochemical reactions occur, and from which secondary biochemical alterations result. It is extremely difficult to interpret the results occurring naturally, for example, on the basis of experiments done *in vitro* with light at 254 nm. In human skin, other ultraviolet spectra and other doses are often of greater interest. Not only do the radiation effects depend upon the optical density of the skin, but also upon light recovery and other factors as well. The photobiologist is, therefore, likely to be most interested in experiments on the human skin made *in vivo*.

One of the immediate reactions of the skin to ultraviolet radiation was discovered by Wels in 1938. He found that ultraviolet radiation increased the reducing power of the skin. This can be replicated *in vitro* by observing the faster reduction of sodium tellurite to black tellurium in the presence of cystein under ultraviolet irradiation (Fig. 5). This effect is abolished

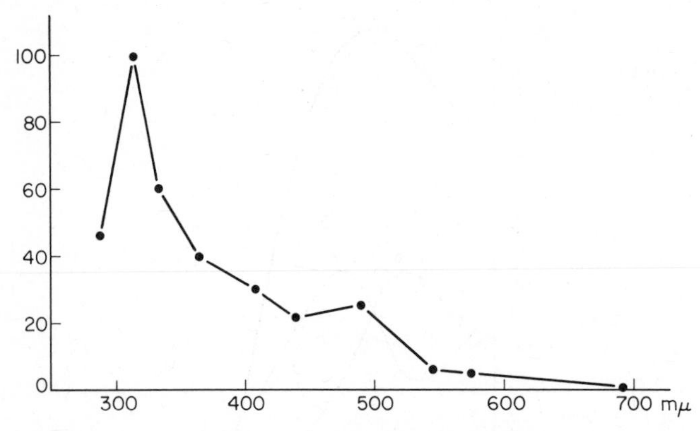

Figure 5. Action spectrum of reduction of a sodium tellurite solution by cystein and ultraviolet light. (Adapted from Spode and Weber, 1953.)

by the use of the sulfhydryl blocking agent, sodium mono-iodo-acetate. The same reaction can be observed *in vivo* in a pig's ear where it is likewise abolished by the sulfhydryl blocking agent. In this reaction middle and long wave ultraviolet light as well as visible light is effective, and the maximum lies at 313 nm. Wels regarded this phenomenon as due to a release of hydrogen from the photo-oxidation of sulfhydryl groups.

Wels's results have recently been confirmed by the experiments of Ogura and Knox. After a single irradiation of human skin with a high pressure mercury lamp, they found a rise of sulfhydryl groups in an epidermal homogenate one and six hours afterwards and a fall below the normal level of sulfhydryl groups from the third to the fifth day. Their experiments revealed a return to normal starting at the tenth day (Fig. 6).

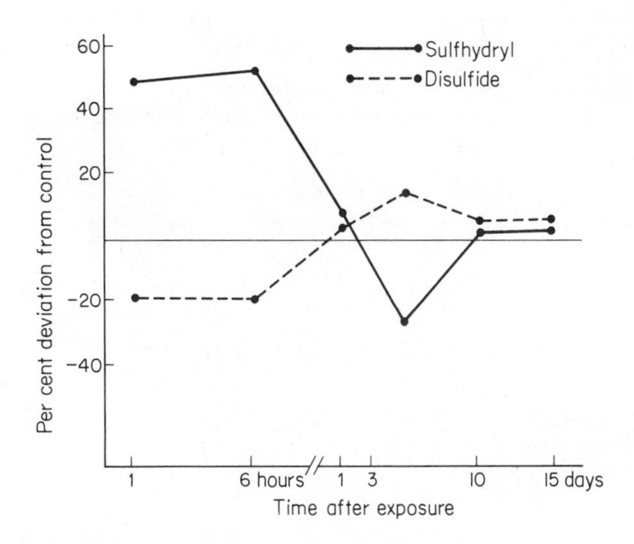

Figure 6. Changes in the sulfhydryl and disulfide concentration following a single exposure to ultraviolet light.

The disulfide groups show the opposite trend.

Wels assumed that the reduction of vital substances was of primary importance. One might also add that various thiol compounds and thiol-containing enzymes such as succinic dehydrogenase would be oxidized and thereby significantly affected.

Histological investigations show particularly well the changes which occur in the epidermal cells. Twenty-four hours after a single radiation of the human skin with the unfiltered mercury vapor lamp and doses lying between four and eight times the minimal erythema dose, distinct signs of cell injury can be seen. These signs include vacuolation of cell cytoplasm, as well as nuclear changes of increased or decreased density.

To decide whether cell injury resulted from the inactivation of DNA, we irradiated the skin with the 254 nm mercury line and with the 302 nm line using a four- to eight-fold minimal erythema dose (Fig. 7). An hour

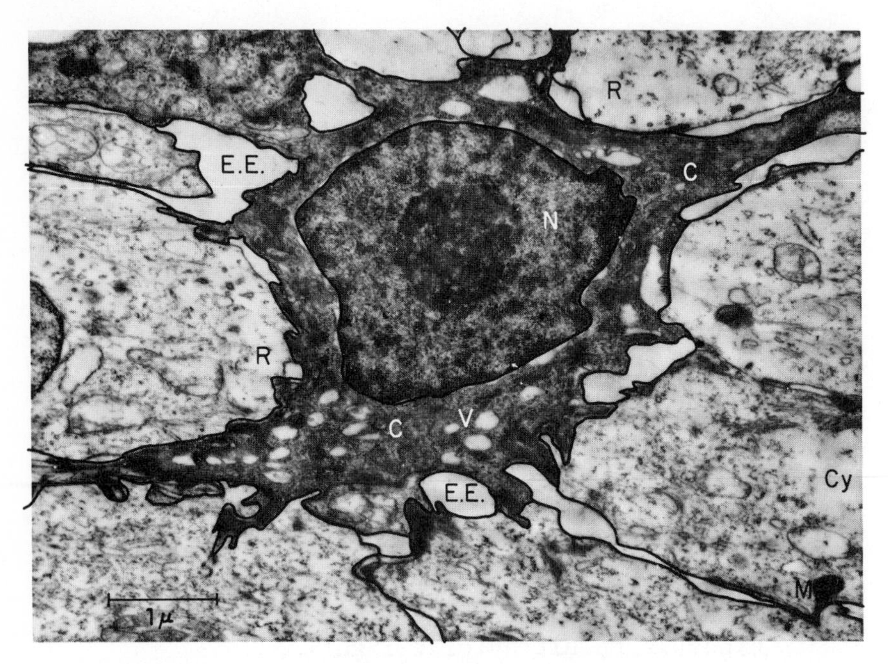

Figure 7. Human epidermis, twenty-four hours after irradiation with 8 MED. A spinous cell shows pyknosis of the nucleus, condensation of cytoplasm, and vacuolation. Peripheral rarification and cytolysis are noticed in other cells. Note also extracellular edema. Abbreviations: C: condensation of cytoplasm; Cy: cytolysis; E.E.: extracellular edema; Me: melanosomes; Mi: mitochondria; N: nucleus; R: rarification of cytoplasm; T: tonofibrils; V: vacuoles.

after irradiation, the nuclei were unaltered. Twenty-four hours later they became hypochromic, as indicated by specific nuclear stains; but no difference in the results could be detected according to the wavelength used. The nucleoli containing RNA were mostly undisturbed even when the nuclei had become hypochromic. Because the nuclear changes took so long to appear, a primary decomposition of DNA by ultraviolet light seemed unlikely. Furthermore, DNA absorbs strongly at 254 nm and has a minimum absorption at 302 nm. So for this reason also, a direct action of the light on DNA seemed unlikely.

The findings under the light microscope are extended by electron microscopy (Figs. 7 and 8). Twenty-four hours after a four to eight times minimal erythema dose, the Malpighian layer may display an intercellular edema and, in some cells, a marginal rarefaction or perinuclear vacuolation of the cytoplasm. We also observe cytolysis and swelling of the nuclei. The later stage of cell destruction is shown by condensation of the cytoplasm and nuclear pyknosis. The irregular dense bodies in human skin described by Nix and his co-workers were not seen by us under the conditions described.

Although Nix and his co-workers spoke of dense bodies, they did not mention lysosomes. The theory of primary lysosomal damage must also be considered. The lysosomes are cell particles whose surrounding membrane may burst under various influences and release hydrolytic enzymes. These enzymes are capable of splitting nearly all cell components, such as proteins and nucleic acids. Lysosomal behavior after ultraviolet irradiation of human skin has been described by Olson *et al.* and by Daniels and his group. Without further investigation, definite conclusions on the importance of lysosomes in ultraviolet injury cannot be offered at present.

Alterations of enzymes are not detectable by histochemical methods before the time at which damage can be seen with the light microscope, namely at twenty-four hours. Increase of glycogen in the basal cells is already visible twelve hours after irradiation. This response is not peculiar to ultraviolet radiation, and it may be seen after one strips the stratum corneum with adhesive tape.

Within a day or a day and a half of the administra-

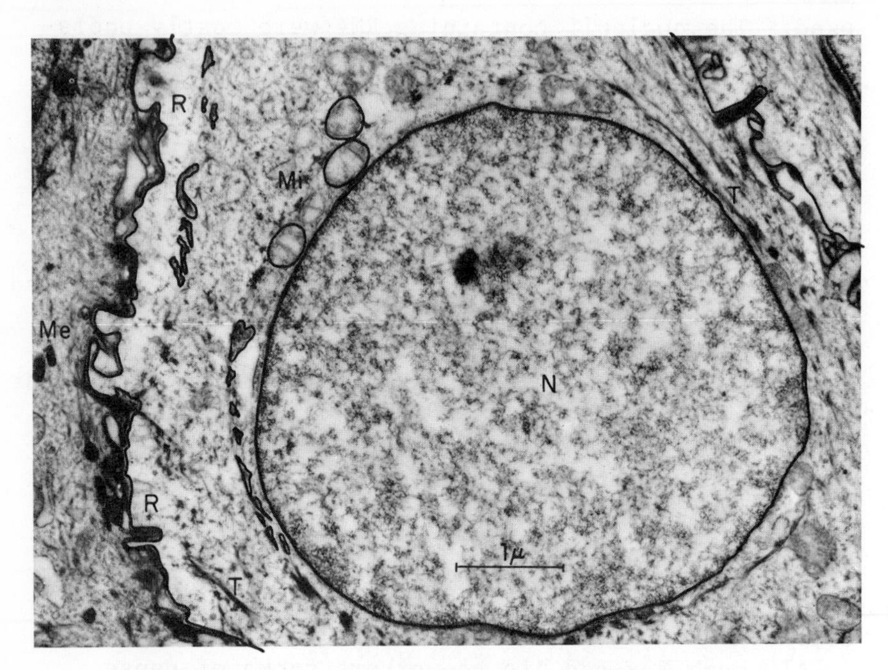

Figure 8. Human epidermis, twenty-four hours after irradiation with 8 MED. A keratinocyte showing swollen nucleus and peripheral rarification of the cytoplasm. (For abbreviations, see Fig. 7.)

tion of weak erythemal doses, signs of increased mitotic activity become manifest (Fig. 9). Soffen and Blum found a gradual increase in changes extending up to a maximum between the eighth and the fourteenth day. These changes consisted of a rise in the mitotic index with increase in nuclear diameter, and an increase in thickness of the epidermis and numbers of epidermal cells per unit area. These changes represent a repair of the cell injury due to ultraviolet light, and at the same time they protect against further ultraviolet irradiation. They correspond to what the clinician calls habituation to ultraviolet light.

Sunburn and suntanning are the visible signs of ultraviolet injury to the skin and the repair of such injury (Fig. 10). Ultraviolet erythema is the inflam-

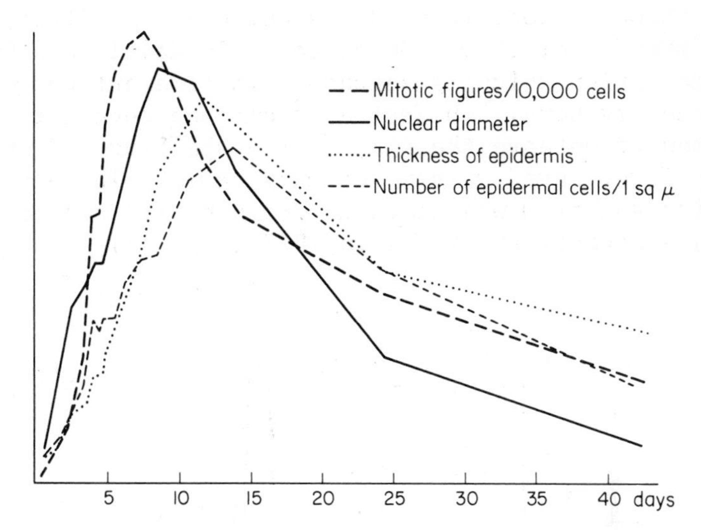

Figure 9. Mouse skin (ears) after single radiation with a mercury intermediate pressure arc, 24.4 ergs X $10^5 cm^{-2}$. (Adapted from Soffen and Blum, 1961.)

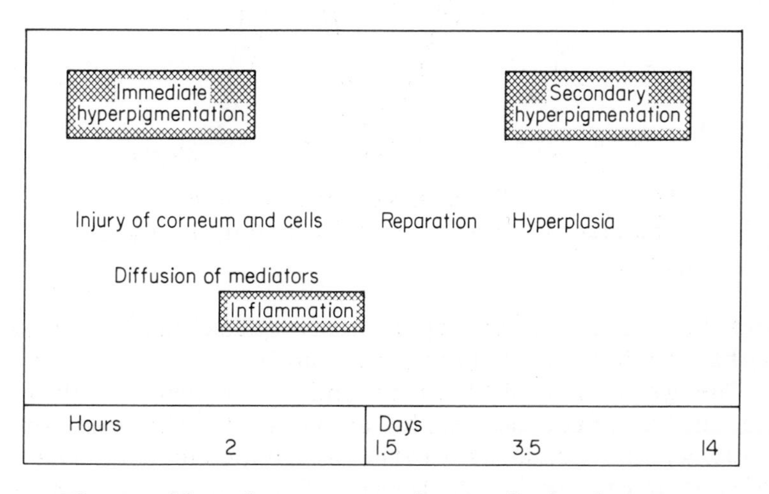

Figure 10. Acute reactions of the human skin after ultraviolet irradiation.

matory reaction to radiation, and it appears after a latent period of a few hours. Three or four days after sun exposure containing doses which are a little below the minimal erythemal dose, the skin develops a hyperpigmentation which is secondary in time but not caused by the erythema. In fact erythema and secondary hyperpigmentation have the same action spectrum and must, therefore, have the same primary cause. The erythemal action spectrum was first determined with a monochromator by Hausser and Vahle in 1927 (Fig. 11). The find-

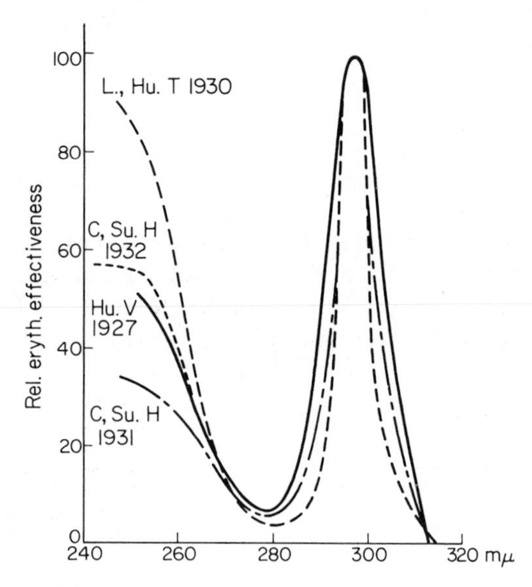

Figure 11. Erythemal action spectra, independently determined by several investigators.

ings have been confirmed on many occasions, most recently by Berger, Urbach, and Davies.

The great variability in the short wave portion of the action spectrum can be ascribed to the time of reading, the grade of erythemal response, and the thickness of the horny layer. The later the reading, the stronger is the grade of erythema. The thicker the horny layer, the weaker is the erythemal effectiveness of short wave ultraviolet light in comparison with medium wave ultraviolet light.

The curve of the erythemal action spectrum may be interpreted in different ways. For instance, it may be said to represent the transmission of stratum corneum and the marked decrease in absorption of almost all biologically important compounds, such as proteins and amino acids, at 300 nm.

Rotter and Van Der Leun have proposed a well-supported hypothesis in favor of two erythemal action spectra produced in different ways. One spectrum has a maximum at 250 nm, while the other has a maximum at 300 nm. Each overlaps the other. According to Van Der Leun, the erythema produced at 250 nm is a result of the direct effect of ultraviolet light on the dermis, while the 300 nm erythema is the result of the diffusion of vasoactive substances from the epidermis to the blood vessels below.

It is widely accepted that ultraviolet light inflammation is produced by the diffusion of mediator substances from the epidermis. The latent period of ultraviolet erythema and the lack of blood vessels in the epidermis support this possibility. Polypeptides (including kinins) and such biogenic amines as histamine, tryptamine, and 5-hydroxy-tryptamine come into consideration. These can be derived from irradiated proteins, free amino acids, and 5-hydroxy-tryptophane, or from cellular decay. Another source of vasodilators may be found in the release of material by degranulation of the mast cells in the dermis.

Histamine as a mediator of radiobiological effects, a concept originally formulated by Lewis and Ellinger, is hardly accepted any longer. Although it has been shown that 5-hydroxy-tryptamine and 5-hydroxy indolacetic acid are excreted in larger amounts in man after ultraviolet radiation, it remains doubtful whether 5-hydroxy-tryptamine could contribute to the genesis of erythema. Kinins are released in tissue by wavelengths in the neighborhood of 300 nm, but not by those around 254 nm. This release occurs during the latent period, and it is not correlated with the appearance of the erythema. Any radiation reaching the dermis should act directly on the capillaries and venules. This is probably what happens with erythema caused by deeply penetrating long wave ultraviolet light, since the erythema develops without a latent period.

Once the ultraviolet inflammation fades, cell proliferation begins. Functioning melanocytes augment and produce more melanin granules (i.e., melanoprotein). These melanin granules are transferred through the dendrites to the keratinocytes in greater amounts and migrate with these cells to the skin surface.

The distribution of melanin granules over all layers of the epidermis produces the optical impression of skin darkening. Fitzpatrick has described this process in detail. The temporal course of erythema and hyperpigmentation may be observed by measuring the reflectance of green and red light from the skin. The demonstrated curves were elaborated by Breit and Kligman (Fig. 12).

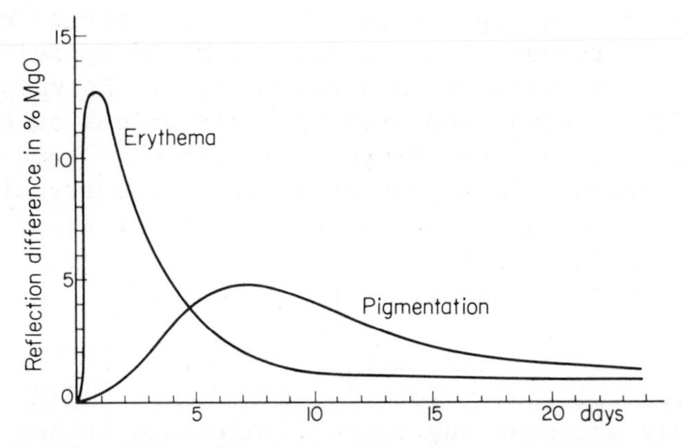

Figure 12. Temporal course of erythema and pigmentation after a 4 MED of ultraviolet light delivered by a hot quartz source. (Adapted from Breit and Kligman.)

Apart from the new formation of melanin pigment, an oxidative hyperpigmentation may be visible immediately after irradiation. By screening off the surrounding skin in the form of some pattern, this pigmentation can be convincingly seen.

The person used in the demonstration was irradiated

by an Osram Xenon high pressure lamp, type XBF 6000, without quartz interposing absorption filters. You will see that immediate pigmentation ends in this case between 500 and 600 nm, according to the absorption filters used.

Pathak *et al.* even found wavelengths above 600 nm and up to 700 nm as being active, with a spectrum at 300 nm (Fig. 13). Henschke and Schulze found that the

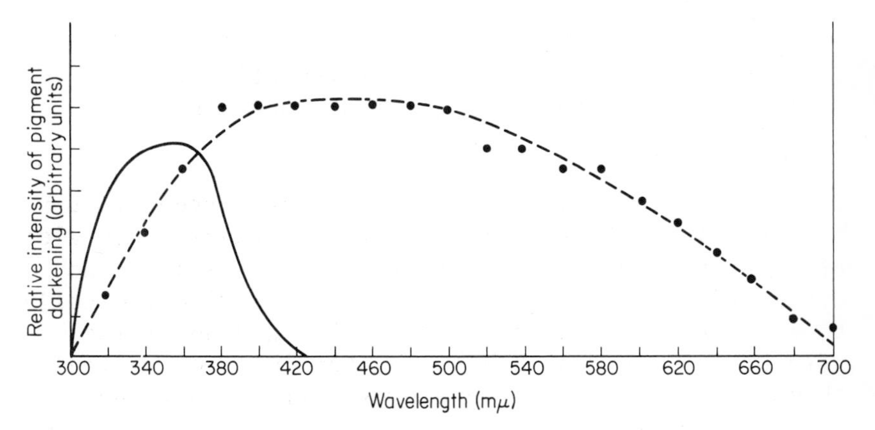

Figure 13. Action spectra for immediate pigment darkening. (Solid line, Henschke and Schulze; broken line, Pathak.)

upper end of the spectrum lay at 450 nm. In my opinion, the discrepancy is due to different dose rates and doses used by these investigators.

The frequency of immediate pigmentation depends on race and skin color as well as age (Fig. 14). In Germans we found a distinct decline in pigmentation at the age of forty-five; the younger people all pigment well, the older only in part.

The phenomenon of immediate pigmentation seems to be due to a photo-oxidation of colorless melanogen which is partly reversible. Pathak was able to associate this process with formation of free radicals. Electron spin resonance can be considerably augmented by irradiation from wavelengths between 320 and 700 nm.

Hyperpigmentation and hyperplasia of the epidermis

Age	Pigmented	Nonpigmented	Total Nr of persons
7–44 years	108	2 (2 %)	110
45–79 years	59	23 (28 %)	82
Total	167	25 (13 %)	192

Figure 14. Relation of pigmentation to age. (Adapted from Wiskemann and Wisser, 1956.)

offer physiological protection against further injury by ultraviolet irradiation. Melanin granules are also of remarkable optical density. Inside the keratinocytes and near the nuclei, they migrate with the cells towards the skin surface, scattering and absorbing part of the incident light. By forming free radicals, they are also able to capture oxidizing and reducing photoproducts.

For a long time, it was believed that thickening of the horny layer was the essential adaptive change to ultraviolet light irradiation in the environment. The idea now prevails that melanin granules in the horny layer and the Malpighian layer accumulate to form a protection which is not only significant but possibly even more important in light protection than in thickening of the horny layer alone.

It is the genetically-fixed, potent power of the individual to become pigmented which regulates the predominant protective mechanism to ultraviolet radiation. Besides the passive light protection by ray filtering, Daniels and Tronnier discuss the habituation to ultraviolet light through the exhaustion of substances which could, by photochemical alteration, be converted into erythematogenic products.

Adaptation allows an increased tolerance to higher and higher doses of ultraviolet light (Fig. 15). By starting with daily exposures to sunlight of 80% of

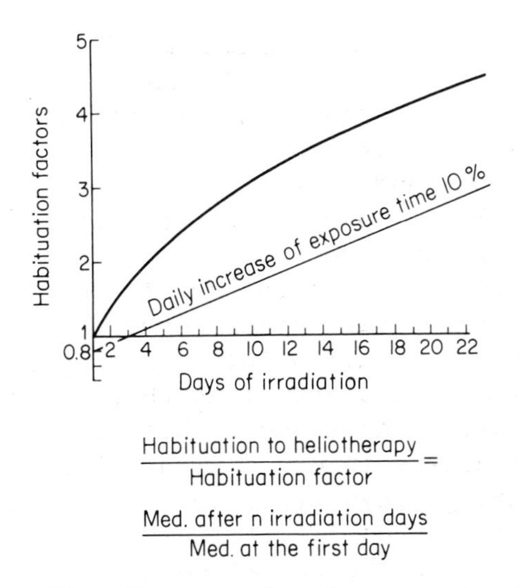

Figure 15. Increased tolerance to irradiation as a result of previous exposure. (Adapted from Schulze, 1952.)

the minimal erythema dose, a 10% increase from one irradiation to the next will compensate exactly for the increased tolerance. It takes one to two months for this tolerance to wear off.

Injury to the epidermis and dermis, brought about by chronic exposure to the sun over many decades, leads to the production of farmers' or sailors' skin (Fig. 16). This skin disease consists of an atrophy,

Epidermis	Dermis
Atrophy	Collagen degeneration
Keratoma	Elastosis
Carcinoma	Furrows

Figure 16. Chronic reactions of the skin after ultraviolet irradiation.

keratoses, and sometimes carcinoma of the epidermis, along with degeneration of collagen and elastic tissue in the dermis. The density and distribution of melanocytes are also altered.

These atrophic and degenerative changes of the visible skin in elderly people, which appear to be due to aging of the tissue, are in fact caused by ultraviolet light. This is also true for senile keratosis, which is perhaps better termed solar keratosis. These chronic skin reactions to ultraviolet light probably follow the identical action spectrum to that of acute sunburn. Experimental skin cancer shows this clearly.

There also exists in man a clear relationship between erythematogenic ultraviolet radiation as determined by the radiation climate, exposure to light, and the skin color, and the development of skin cancer. Urbach and his co-workers found that the localization of frequencies of squamous carcinoma of the skin on the head corresponded with those areas exposed to the maximum ultraviolet light. The sites of occurrences of basal cell carcinomas are not in full correspondence with the areas of maximum radiation.

In this view, a great deal has been said about the harmful effects of sunlight. Nevertheless, millions of people are eager to spend all their weekends and holidays in search of sunshine.

On the credit side, it certainly is possible to have too little ultraviolet light. Symptoms of rickets may arise, and some skin diseases such as psoriasis are actually promoted by reduced light. Ultraviolet radiation stimulates reproductive power and may prevent catarrhal diseases. These and other effects of ultraviolet light on the body as a whole are probably regulated by the neurovascular supply of the skin via the diencephalon and the pituitary. In this way, solar radiation may influence the autonomic functions and the endocrine glands. These changes lead to immunological and metabolic benefits. So we see that ultraviolet light on the skin produces both injury and benefit to human beings.

Photomorphogenesis

H. Mohr

University of Freiburg

By the term "photomorphogenesis" we designate the control which may be exerted by light over growth, development, and differentiation of a plant, independently of photosynthesis. Figures 1 and 2 illustrate some characteristic phenomena of photomorphogenesis. The two mustard seedlings (Fig. 1) have virtually the same genes, and both were grown on the same agar medium which contained an abundance of organic molecules. "Light" must be responsible for the different appearance. The dark-grown seedling on the right shows the characteristics of "etiolation," whereas the light-grown seedling has developed in a way which we designate by the term "normal." It is easy to demonstrate that the influence of light is not due to photosynthesis. We have rather to conclude that in a seedling some photochemical reaction must occur independently of photosynthesis in order to make possible what we

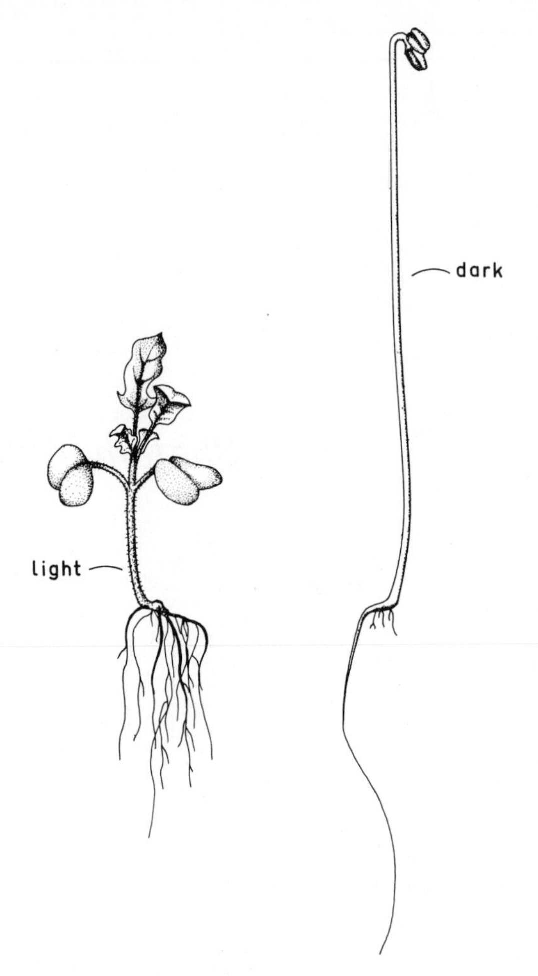

Figure 1. These two mustard seedlings (*Sinapis alba* L.) have the same chronological age and are virtually genetically identical. The differences in morphogenesis are due to the light.

are used to calling normal growth and development. This statement is probably true for all higher plants. A second example is shown in Fig. 2. The sprouts of a potato tuber will etiolate in darkness whereas in the light, the normal potato plant will develop. A characteristic of the etiolation is that the internodes grow

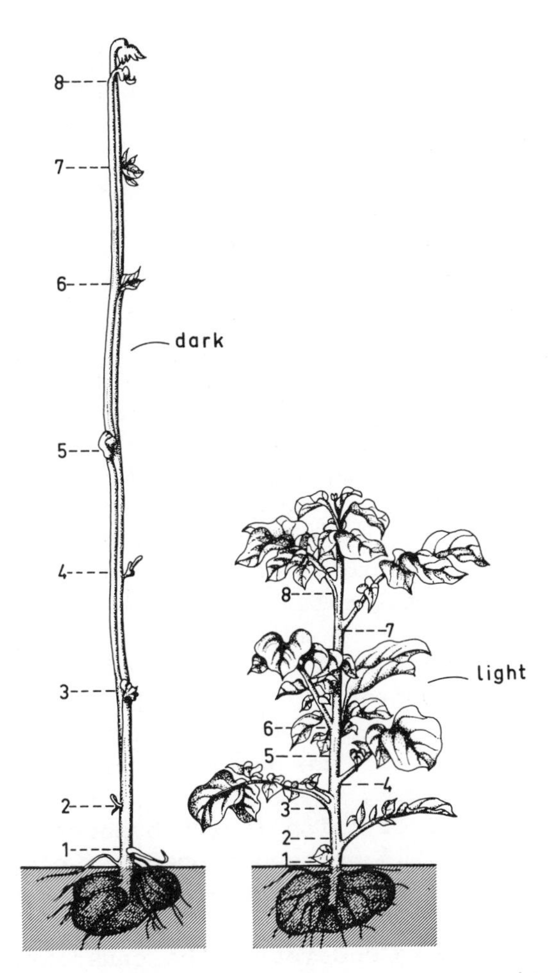

Figure 2. These two potato plants (*Solanum tuberosum* L.) are genetically identical. Nevertheless, the dark-grown and the light-grown plants differ greatly.

rapidly while the leaves remain rudimentary. The biological significance of etiolation is obvious. The physiological problem is to understand in a modern, preferentially "molecular" terminology how light can force a young plant to develop in a "normal" instead of an "etiolated" manner. Obviously two things are involved: first, a photochemical reaction; and second, a change in the manner in which the genes of the particular organism are used. We recognize that the causal analysis of photomorphogenesis may contribute to our knowledge of development and lead to a better understanding of a basic problem of biology, control of differentiation and development in multicellular systems.

Seedlings of higher plants (e.g., mustard or lettuce seedlings) proved to be useful systems for the causal analysis of photomorphogenesis. The first stages of vegetative growth after seed germination (Fig. 3) are especially suitable for the following reasons. First, the seedling consists during this phase of only three parts: cotyledons, hypocotyl, and radicle. The plumule is hardly developed. Second, the seedling contains so much storage material (mainly fat and protein) in the cotyledons that it is completely independent of photosynthesis, or of the external supply of organic molecules, for a number of days after germination. Third, the seedling can be grown on a medium which supplies only water. An external supply of ions is not required. The seedling can therefore be regarded as a closed system for nitrogen or phosphate. Fourth, during the period of experimentation which we normally use, there is no significant DNA increase and no cell division in the hypocotyl nor in the cotyledons. This photomorphogenesis can be studied without getting involved in problems of DNA replication or changing cell numbers.

The problem is to understand the causalities of photomorphogenesis. The developmental biologist wants eventually to describe in a molecular terminology the total pathway which leads from the photochemical event to the final photoresponse as represented by flowering, cell growth, the synthesis of anthocyanin, or the formation of a plastid. In order to approach this problem, we ask the following questions: what type of pigment absorbs the effective light; what kinds of photochemical reactions occur; and what are the causal relationships between the photochemical reactions and the final photoresponses which we can plainly see and measure.

And the final question is how can the integration of the photoresponses of the different cells, tissues, and

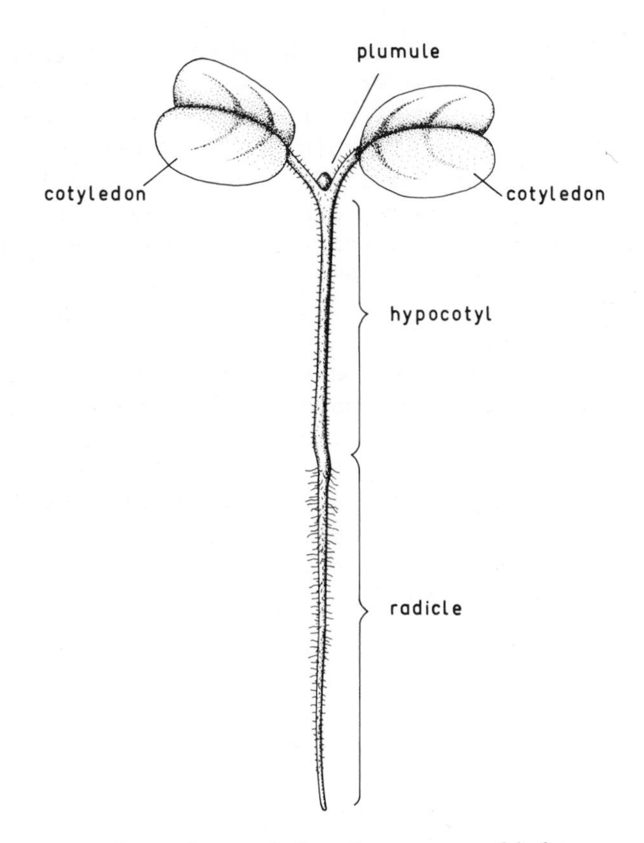

Figure 3. The model of a young light-grown mustard plant. During the first stage of vegetative growth after seed germination, the seedling virtually consists of only three parts: cotyledons, hypocotyl, and radicle.

organs be understood? Photomorphogenesis is a complex but harmonic process, precisely regulated in space and time. What I have outlined is a vast research program which includes virtually all important topics of developmental biology. It seems wise to deal with the problem step by step.

In the first step we ask: What type of pigment absorbs the effective light? The answer is phytochrome (at least above 520 nm). I shall restrict this article to a discussion of phytochrome for two reasons. Phytochrome is the only photomorphogenic photoreceptor which has been isolated and studied *in vitro*, and only in the case of phytochrome has photobiology been successfully related to the molecular biology of development. Most biologists are familiar, of course, with at least some of the established knowledge concerning phytochrome. I shall mention only a few results of phytochrome research in order to create some basis for the further question we shall discuss in more detail: How does phytochrome act?

Phytochrome is widely distributed in the plant kingdom. It is apparently found in all higher plants, in green algae, and in red algae. Phytochrome is a bluish biliprotein having two interconvertible forms: phytochrome 660, or P_r, with an absorption maximum in the red around 660 nm; and phytochrome 730, or P_{fr}, with an absorption maximum in the far-red around 730 nm (Fig. 4). P_{fr} is thought to be the physiologically

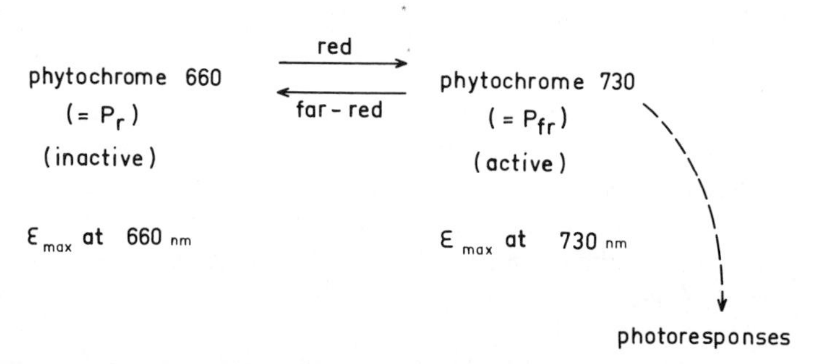

Figure 4. A simple formulation for the phytochrome system.

active form of the phytochrome system, the effector molecule. I agree with this widely-accepted statement, but I must correct it insofar as nobody has shown that P_{fr} can act without the presence of P_r. Therefore, the hypothesis is still possible that both forms of phytochrome, P_r and P_{fr}, are required for phytochrome action. The statement, however, that without P_{fr} no photomorphogenesis can take place, is fully justified. Phytochrome must be supposed to initiate basic changes in the metabolism and in the energetics of the cells and tissues leading finally to the photoresponses (e.g., growth of a leaf or formation of compounds like anthocyanins). In a dark-grown seedling only P_r will be present. Both the phototransformation of P_r to P_{fr} and the reverse reaction involve a photochemical step and a number of dark reactions. A speculated structure proposed for the phytochrome chromophore by Dr. Siegelman is shown in Fig. 5. The pigment

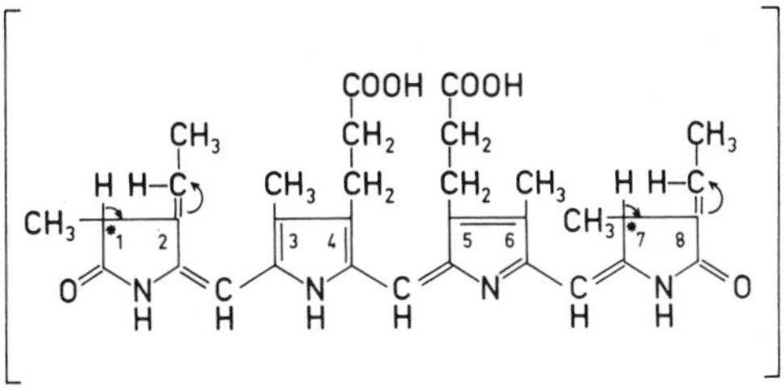

Figure 5. The structure recently proposed for the phytochrome chromophore by Dr. H. W. Siegelman.

is a bilitriene. The arrows indicate how the increase from 7 to 10 conjugated double bonds during the $P_r \rightarrow P_{fr}$ transformation might occur. The molecule contains two asymmetric centers which would be apparently lost on transformation to the longer wavelength's absorbing form. The transformation of the 660 nm-absorbing form of phytochrome to the 730 nm-absorbing form is a multistep process, and the model allows for several intermediates considering both electron shifts and hydrogen migration. The phototransformations of phytochrome occur only when the chromophore is attached to its protein. However, the molecular basis of the cooperative interaction between the protein and the chromophore is not yet known. A phytochrome model recently published by Dr. Crespi and associates accounts at least for conformational changes of the protein. They propose that the chromophore is covalently attached to the peptide backbone through an ethylidene group. Conversions between P_r and P_{fr} would lead to backbone reflexes that would manifest themselves in conformational intraconversions in the protein.

The absorption spectra of P_r and P_{fr} overlap over a wide range (< 740 nm). Under conditions of saturating irradiations, this overlap causes the phytochrome system in the cell to be characterized by photostationary states (Fig. 6). We recognize that, at the most,

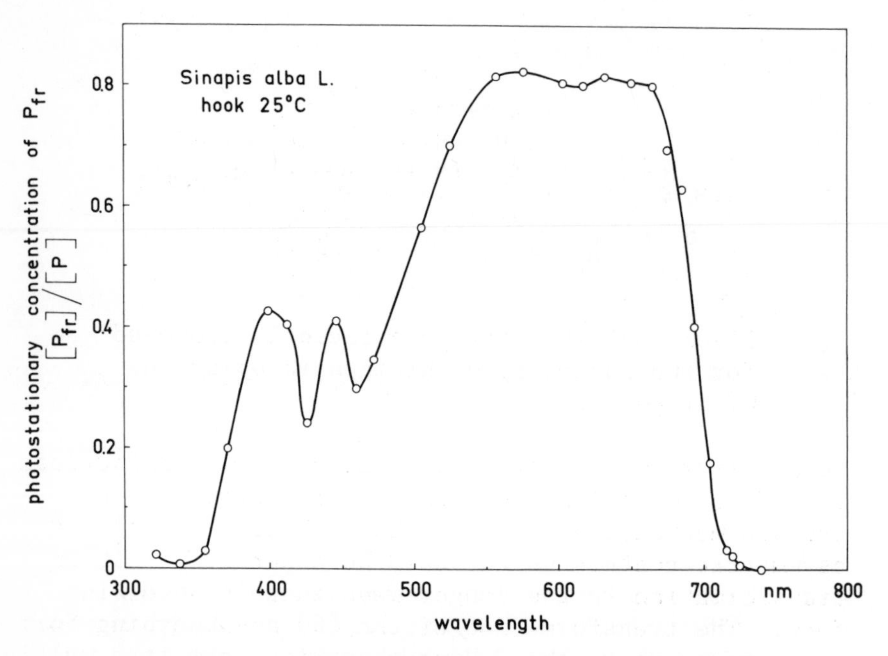

Figure 6. The percentage of P_{fr} at photoequilibrium as a function of wavelength (*in vivo* measurements at 25°C with mustard seedlings, hook region). (From experiments of Dr. K. M. Hartmann and Dr. C. J. P. Spruit.)

about 80% of the total phytochrome can be present as P_{fr}, for instance, if we irradiate the living system with pure red light around 660 nm. If we irradiate with pure far-red around 720 nm, only about 3% of the total phytochrome will be present as P_{fr}. The photostationary state of the phytochrome system is rapidly established, at least in the red and far-red where the extinction coefficients of the phytochromes are high. A few minutes of irradiation with medium quantum flux density are satisfactory for establishing the photostationary state. In brief, only "short-time irradiations" are required to establish photostationary states in the phytochrome system *in vivo* as well as in the photochemically active extract.

This fact makes understandable the well-known induction-reversion experiments which are often used as a criterion of the involvement of phytochrome (Fig. 7).

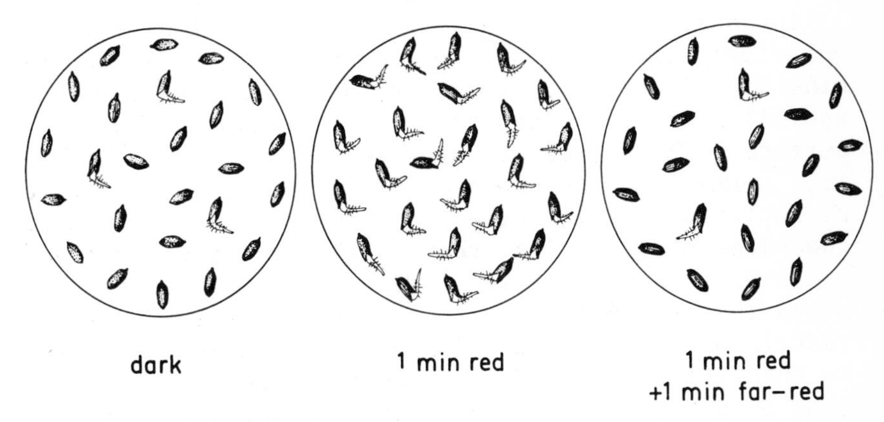

Figure 7. Reversion of an induction by red light of lettuce seed germination by a subsequent irradiation with far-red light.

If lettuce seeds of a light-sensitive variety are sown on a suitable medium at 25°C and placed in darkness, most of the seeds do not germinate. One minute of red light induces 100% germination. If we follow the red with far-red light, the induction of germination is nullified. The explanation is that the red treatment establishes a photostationary state where about 80% of the total phytochrome is present as P_{fr}. Under these conditions all seeds start to germinate. The far-red treatment, on the other hand, leads to a photostationary state where only a few percent of the total phytochrome is present as P_{fr}. Under these conditions germination cannot proceed since the P_{fr} requirement is not satisfied.

Where is phytochrome located inside the cell? There is at least one piece of experimental evidence. With the help of linearly polarized red and far-red light, it could be shown in physiological experiments with several types of cells that molecules of phytochrome are oriented in a dichroic structure in the outer cytoplasm, possibly along the plasmalemma. For example, the phenomena of phytochrome-mediated polarotropism can only be understood if we assume that phytochrome molecules are strictly oriented in the cell.

The basic observation (Fig. 8) is that linearly polarized light applied from above can strictly determine the direction of growth of a filamentous fern sporeling growing under special conditions on an agar surface. The sporelings grow at an angle of 90°, that is, "normal" to the plane of vibration of the electrical vector of the linearly polarized light. If we turn the plane of vibration, the sporelings rapidly change their direction of growth. We know from other experiments that the growth of this filamentous system is restricted to the upper 10 μ of the filament, that is, to the extreme tip. The action spectrum indicates that above 500 nm, the polarotropic response is mediated by phytochrome. The further results obtained with this system can only be understood, however, if one assumes that the axis of maximum absorption of the phytochrome molecules turns by 90° during the transition of the red absorbing form P_r to the far-red absorbing form P_{fr} leading to an orientation of the P_{fr} molecules perpendicular to the cell surface (Fig. 9). This concept has recently been confirmed for the green alga *Mougeotia* by Dr. Haupt. He concluded from experiments on phytochrome-mediated chloroplast movement, that the surface-parallel orientation is restricted to the P_r form, whereas the P_{fr} form must be assumed to be orien-

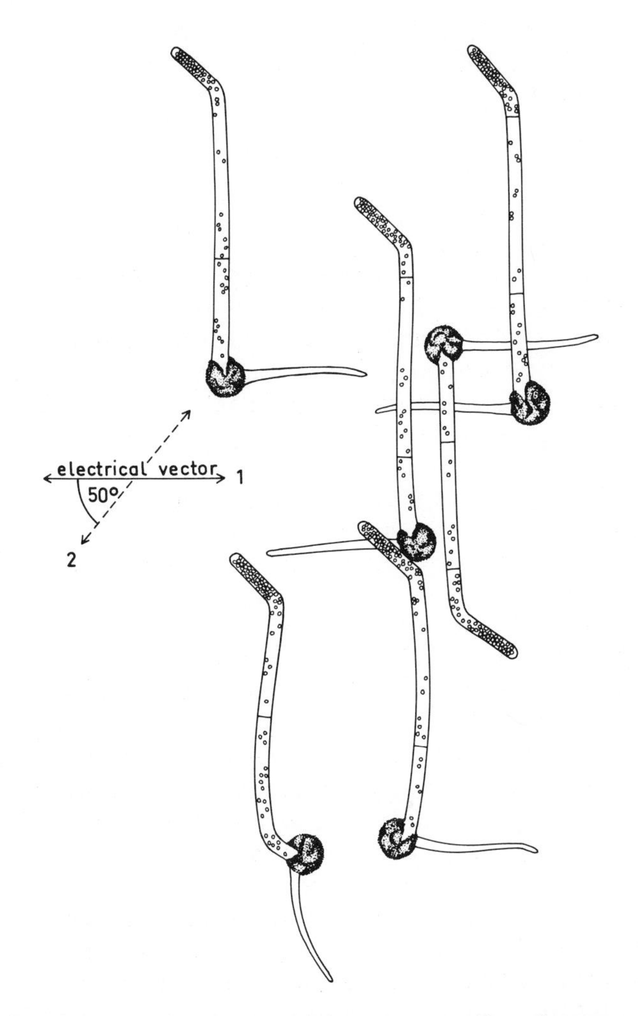

Figure 8. A basic phenomenon of polarotropism. Linearly polarized light applied from above can strictly determine the direction of growth of a fern protonema (filamentous fern sporeling). The sporelings grow normally to the plane of vibration of the electrical vector.

ted normally to the surface.

It should be emphasized, perhaps, that there is no reason to assume that phytochrome molecules are always

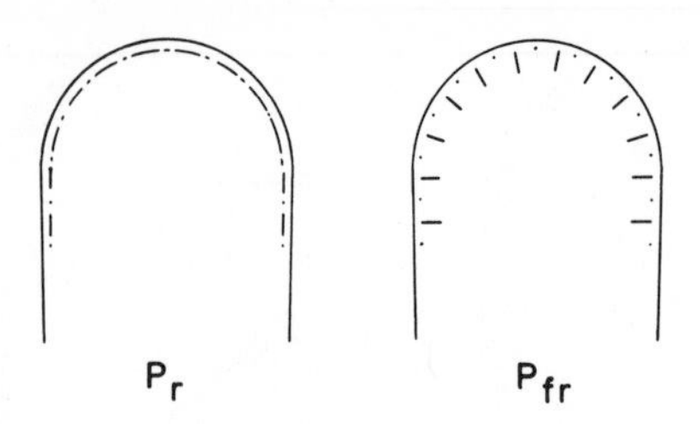

Figure 9. This model (tip of the fern protonema) may illustrate the orientation of the axis of maximum absorption of the phytochrome chromophore. Left: only P_r present. Right: part of phytochrome present as P_{fr}. (After Etzold, 1965.)

and exclusively oriented in the manner just described. We rather prefer the concept that even within the same cell phytochrome molecules may be located in different "compartments."

Before we close the chapter on phytochrome properties, we have to mention one further characteristic of far-reaching importance. Whereas P_r is stable in darkness, P_{fr}, the physiologically active species of phytochrome, is not. It can disappear in two ways: it is readily destroyed by an irreversible reaction or it reverts slowly to P_r in a thermal reaction. In an etiolated seedling the irreversible destruction of P_{fr} seems to be the dominant process. This is a very critical point for in order to study the action of P_{fr} at the molecular level, it seems advisable to run long-time kinetic studies under conditions in which P_r is rate-limiting and the concentration of the effector molecule P_{fr} remains stationary over an extended period of time. How can this be achieved in view of the fact that P_{fr} is not stable? The problem can be solved by the use of continuous far-red irradiation.

Figure 10 shows an action spectrum for control of hypocotyl lengthening in lettuce seedlings under continuous long-time irradiation. If we exclude the blue part of the spectrum from our considerations, we recognize that quanta between 520 and 700 nm are hardly effective and that far-red above 750 nm is ineffective.

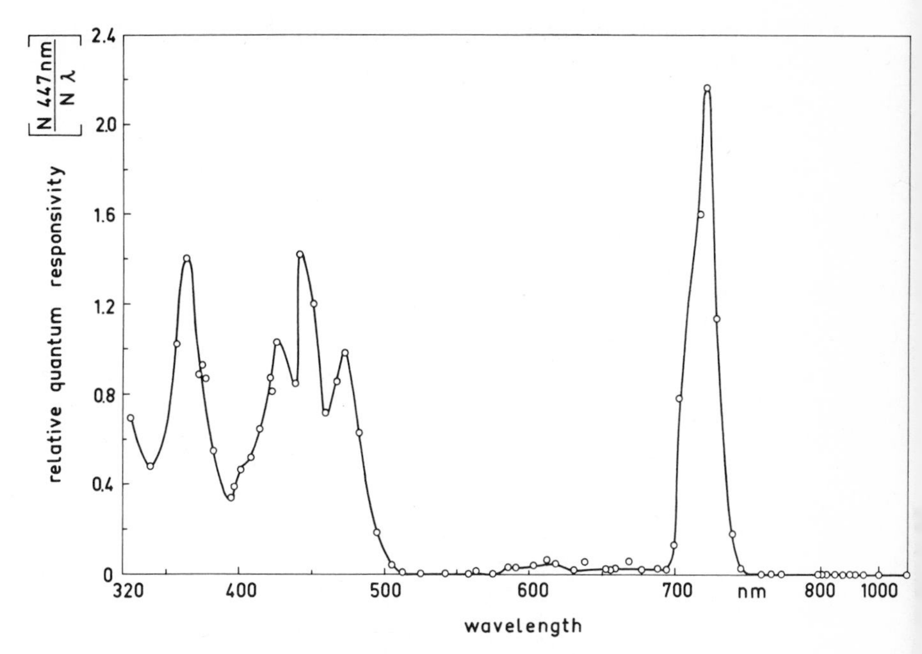

Figure 10. An action spectrum for control of hypocotyl lengthening by light in lettuce seedlings (*Lactuca sativa* L., cv. Grand Rapids). The action spectrum was elaborated between 54 and 72 hr after sowing. During this period lengthening of the hypocotyl is almost exclusively due to cell lengthening. (Supplemented after Hartmann, 1967.)

By contrast, we observe in the near far-red a symmetric band of action close to 720 nm, where the photoequilibrium of the phytochrome system contains about 3 percent P_{fr}. There is convincing evidence, elaborated by Dr. Hartmann, that this far-red band of action can be fully attributed to phytochrome. The experimental evidence which cannot be described further here can only be understood if we assume that under conditions of long-time irradiation, phytochrome is most effective if the ratio of P_{fr} to total phytochrome is low (i.e., on the order of a few percent). Continuous far-red (e.g., a band centered around 720 nm) maintains this sort of photostationary state over an extended period of time (Fig. 11).

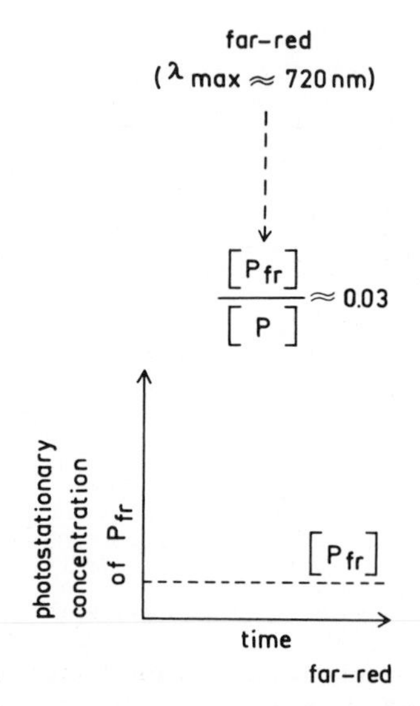

Figure 11. This model illustrates the conclusion that continuous far-red light (e. g., a band centered around 720 nm) maintains a relatively low but virtually constant photostationary concentration of P_{fr} over an extended period of time.

A serious problem was posed by the fact that the photomorphogenic effect of a given far-red source is a function of intensity, whereas the photoequilibrium of the phytochrome system at a particular wavelength does not depend on intensity. At least at higher intensities, the influence of the P_{fr} decay on the photoequilibrium can be ignored. Dr. Hartmann has provided experimental evidence that not only the particular photoequilibrium, but also the total absorption of quanta in the phytochrome system (i.e., the rate of phytochrome cycling), is essential for the physiological effectiveness.

This explanation of the effects of long-time irradiation on the basis of phytochrome is consistent with recent direct measurement of total phytochrome in axis tissue of dicotyledonous seedlings. Drs. Clarkson and Hillman found that in tissue of dark-grown seedlings, the phytochrome content remains constant under continuous far-red irradiation for periods of many hours. We have obtained the following data for the mustard seedling hypocotyl (Fig. 12). Under continuous standard

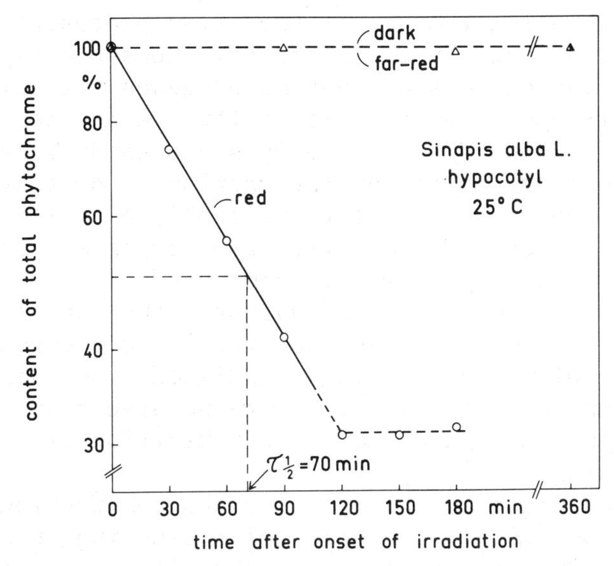

Figure 12. Changes of total phytochrome under continuous red or far-red light. The material is a hypocotyl tissue of *Sinapis alba* L. Measurements were performed by Drs. P. Schopfer and D. Marmé using a "Ratiospect" instrument. The signals Δ (Δ O.D.) represent the total phytochrome content of the sample.

far-red, the phytochrome content as measured by Δ (Δ O. D.) remains virtually constant over a considerable period of time; whereas under continuous red, even of low intensity, the phytochrome content falls rapidly at a constant decay rate until a stationary level is approached rather abruptly. I shall adhere in the following (without any further documentation) to the concept (Fig. 11) that photomorphogenesis of the dark-grown lettuce or mustard seedling in the wavelength range above 520 nm is due to the formation and maintenance of P_{fr}. Under continuous far-red a low, but highly effective and virtually stationary, concentration of the active P_{fr} can be maintained over a considerable period of time. The photostationary state is established within minutes after the onset of far-red. The effectiveness of the given photostationary state is a function of intensity. Consequently, if we analyze the process of photomorphogenesis under continuous far-red of constant intensity, we are dealing from the very beginning with steady state conditions of P_{fr}. Furthermore we avoid any interaction of photosynthesis since, under far-red light, virtually no chlorophyll will be formed. The plastids are completely inactive as far as photosynthesis is concerned.

The next step is to ask what are the causal relationships between P_{fr} and the final photoresponses, some of which might be light-mediated growth of the cotyledons of a seedling, light-mediated synthesis of anthocyanin, or light-mediated inhibition of hypocotyl lengthening.

The experts presumably agree that phytochrome might be the same in all the cells of a seedling in which it occurs. But, as we all know, the different organs and tissues of the seedling respond differently to the formation of P_{fr}. Some of the many photoresponses of the mustard seedling are enumerated in Table 1. We cannot here discuss the multiplicity of displays. I might draw the following conclusions: The response which takes place apparently depends on the "specific status of differentiation" of the cells and tissues. The fate of any particular cell following the formation of P_{fr} depends on its previous history. It determines the *specificity* of the response. P_{fr} can only trigger the response. It has nothing to do with the *specificity* of the response. This point of view is emphasized in Fig. 13. We look at segments of cross sections through the hypocotyl of mustard seedlings. Under the influence

TABLE 1

Photoresponses of the mustard seedling, *Sinapis alba* L. (Supplemented after Mohr, 1966)

Inhibition of hypocotyl lengthening
Inhibition of translocation from the cotyledons
Enlargement of cotyledons
Unfolding of lamina of the cotyledons
Hair formation along the hypocotyl
Opening of the hypocotylar ("plumular") hook
Formation of leaf primordia
Differentiation of primary leaves
Increase of negative geotropic reactivity of the hypocotyl
Formation of tracheary elements
Differentiation of stomata in the epidermis of the cotyledons
Formation of plastids in the mesophyll of the cotyledons
Changes in the rate of cell respiration
Synthesis of anthocyanin
Increase in the rate of ascorbic acid synthesis
Increase in the rate of chlorophyll accumulation
Increase of RNA synthesis in the cotyledons
Increase of protein synthesis in the cotyledons
Changes in the rate of degradation of storage fat
Changes in the rate of degradation of storage protein

of P_{fr} certain cells of the epidermis have formed hairs, and all the cells of the subepidermal layer (but no other cells) have formed anthocyanin. The initial lag-phase after the onset of light is, in both responses, about 3 hr. It is obvious from these simple drawings that P_{fr} functions only as a "trigger." The specificity of the cellular response (e.g., hair formation or anthocyanin synthesis) must depend on the specific status of differentiation of the cells and tissues at the moment when P_{fr} is formed in the seedling.

To approach the problem of phytochrome action further, it was useful to divide the multiplicity of displays of the mustard seedling into three distinct cate-

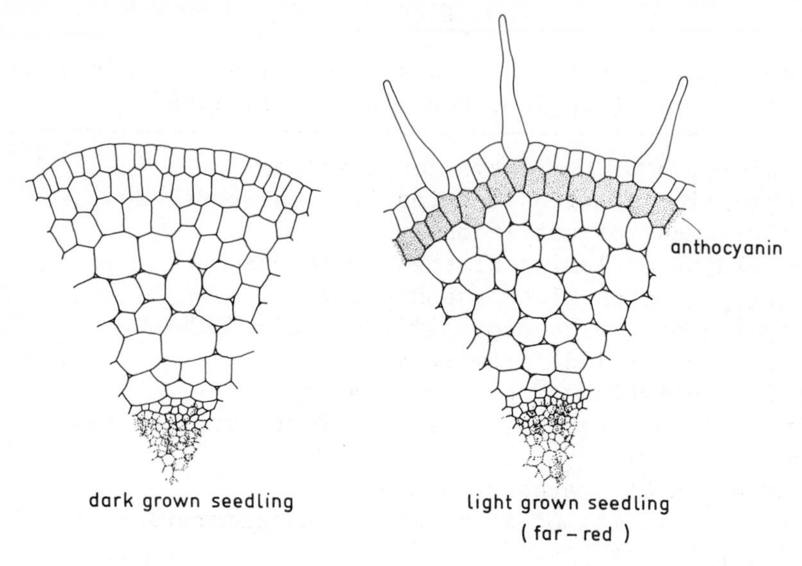

Figure 13. Segments of cross sections through the hypocotyl of mustard seedlings (*Sinapis alba* L.) grown in the dark under far-red light.

gories: positive, negative, and complex photoresponses. "Positive" photoresponses are those which are characterized by an initiation or an increase of biosynthetic or growth processes. An example is phytochrome-mediated synthesis of anthocyanin (Fig. 14). The nearly three hr lag-phase after the first onset of light is characteristic of this type of a photoresponse. (Remember that continuous far-red is supposed to maintain in the seedling a low but stationary concentration of P_{fr} over a considerable length of time, and that the photostationary state is established within minutes after the onset of far-red.) "Negative" photoresponses are those which are characterized by an inhibition of growth processes or other physiological processes, such as translocation. Inhibition of hypocotyl lengthening is a typical response of this sort (Fig. 15). A characteristic feature of a negative photoresponse is the short lag-phase. In the present case the extrapolation shows that within 15 min after the onset of far-red a new, steady state

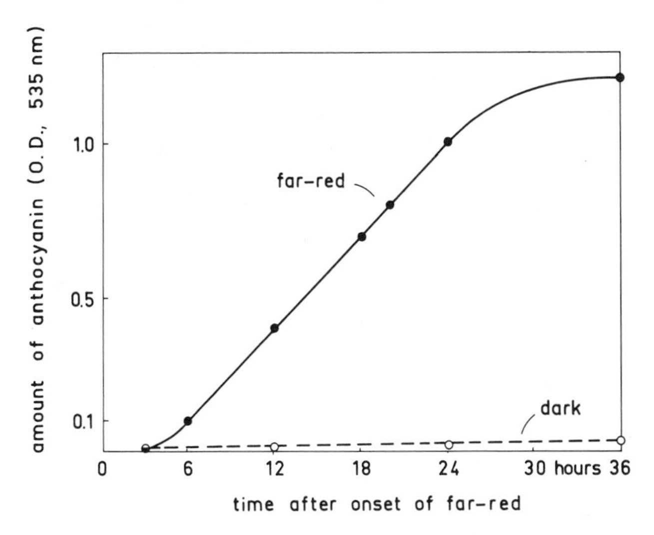

Figure 14. Time course of anthocyanin accumulation in mustard seedlings under continuous standard far-red. The lag-phase of the response is about 3 hr. Onset of far-red: 36 hr after sowing. (After Mohr, 1966.)

rate of lengthening is established. "Complex" photoresponses are those (Fig. 16) which are characterized during the first part of the kinetics by an inhibition and later by a promotion of the response as compared with that of the corresponding dark controls. The control by phytochrome of the rate of O_2 uptake of the cotyledons is an example of this kind of photoresponse. Complex photoresponses can possibly be explained by an interaction of positive and negative photoresponses. As a matter of fact, we are left with positive and negative photoresponses which are different in all respects except that both types are mediated by the effector molecule P_{fr}.

We now again raise the question: How does P_{fr} mediate the photoresponses? We shall restrict this article mainly to those photoresponses of the mustard seedling which lend themselves relatively easily to an analysis in terms of *molecular* events. We shall briefly deal

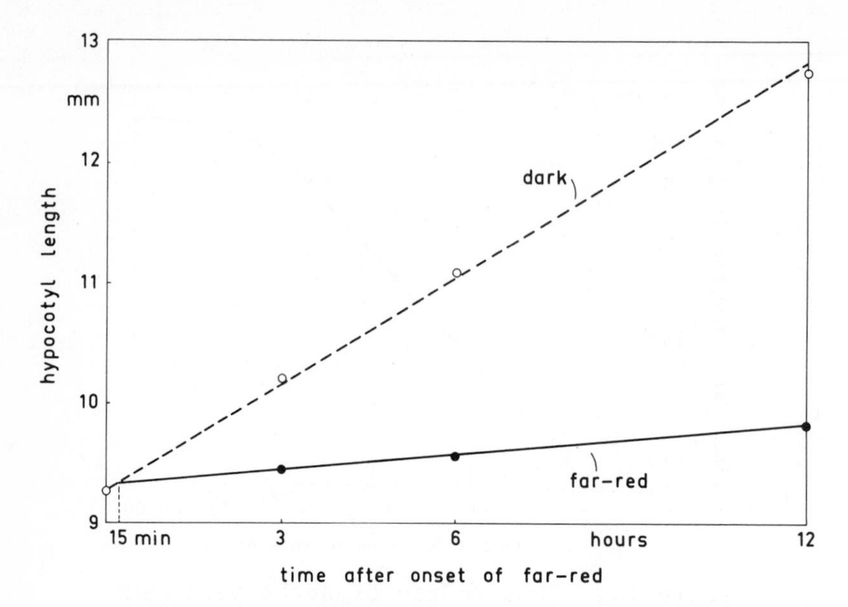

Figure 15. Inhibition of hypocotyl lengthening of the mustard seedling by continuous standard far-red. The lag-phase of the response is short. Onset of far-red: 36 hr after sowing. (After experiments by K. Roth and W. Link.)

with each of the following topics: control of plastid formation by phytochrome, control of anthocyanin synthesis by phytochrome, and control of enzyme synthesis by phytochrome.

The first example is control of plastid formation by P_{fr}. Figure 17 shows the kinetics of protein levels in attached cotyledons of the mustard seedling in darkness and under the influence of far-red, that is, under the influence of a low but virtually constant concentration of P_{fr}. In order to understand these two curves, we must realize that the cotyledons of the mustard seedling contain much storage protein which will be degraded during the development of the seedling. The resulting soluble compounds are either translocated to other parts of the seedling or used within the cotyledons for the synthesis of enzymes and structural proteins. This accounts for the strong relative increase

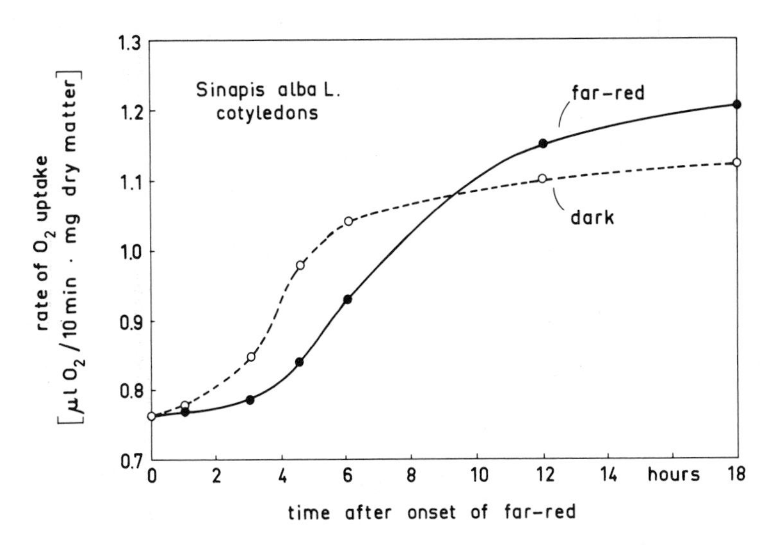

Figure 16. Control by continuous far-red of the rate of O_2 uptake by the cotyledons of the mustard seedling. Onset of far-red: 36 hr after sowing. (After Mohr, 1966.)

in protein under the influence of far-red. Histochemical studies have shown that, under the influence of far-red, the degradation of storage protein in the cotyledons is enhanced. This fact is indicated, in principle at least, in the central part and on the right side of Fig. 18. We are looking at mesophyll cells of mustard cotyledons. The dense dark bodies represent storage protein 72 hr after sowing. It is evident from Fig. 18, and it has been measured quantitatively, that far-red enhances the degradation of the storage protein. At the same time, however, P_{fr} will stimulate a strong *de novo* synthesis of structural protein in the cotyledon. This fact is indicated on the right side of Fig. 18. The lens-shaped, large, proteinaceous bodies which appear under the influence of far-red we call "plastids." As far as size and shape are concerned, they are indistinguishable under the light microscope from chloroplasts except that they do not contain chlorophyll. The longest diameter of the plastids is about 5 μ, whereas that of the corre-

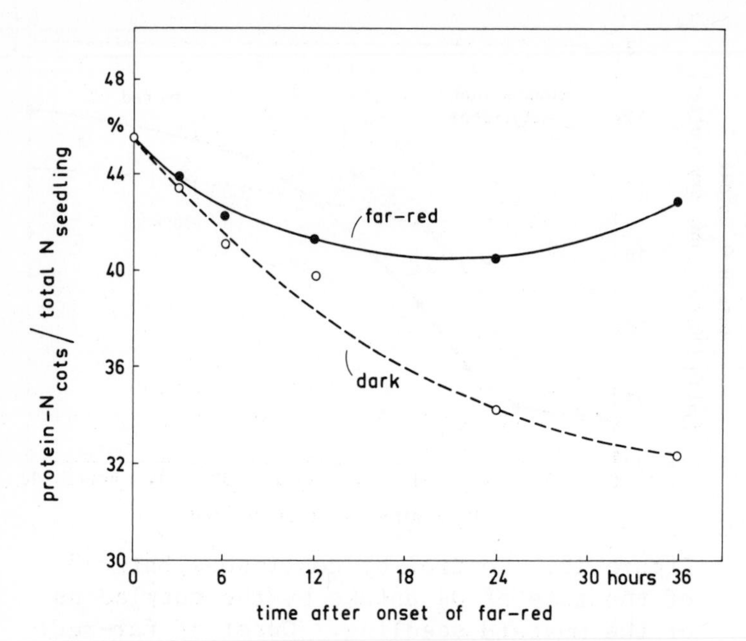

Figure 17. Changes of protein contents in attached cotyledons of the mustard seedling in darkness and under continuous standard far-red. (Remember that the seedling is a closed system for nitrogen over the whole period of experimentation. Total N per seedling does not change.) (After Jakobs and Mohr, 1966.)

sponding etioplasts is only about 1.5 μ. "Normal" chloroplasts formed under white light in the cotyledons are shown on the left side of Fig. 18. We conclude that growth and development of plastids is to a large extent under the control of phytochrome.

The electron microscope reveals that 72 hr after sowing, the dark-grown seedling contains only small etioplasts each of which has a prolammelar body and very few thylakoids. Treatment with white light leads to the formation of normal cotyledon chloroplasts. Treatment with far-red leads to the formation of plastids which have the same size as those in white light, but which possess an internal structure similar to

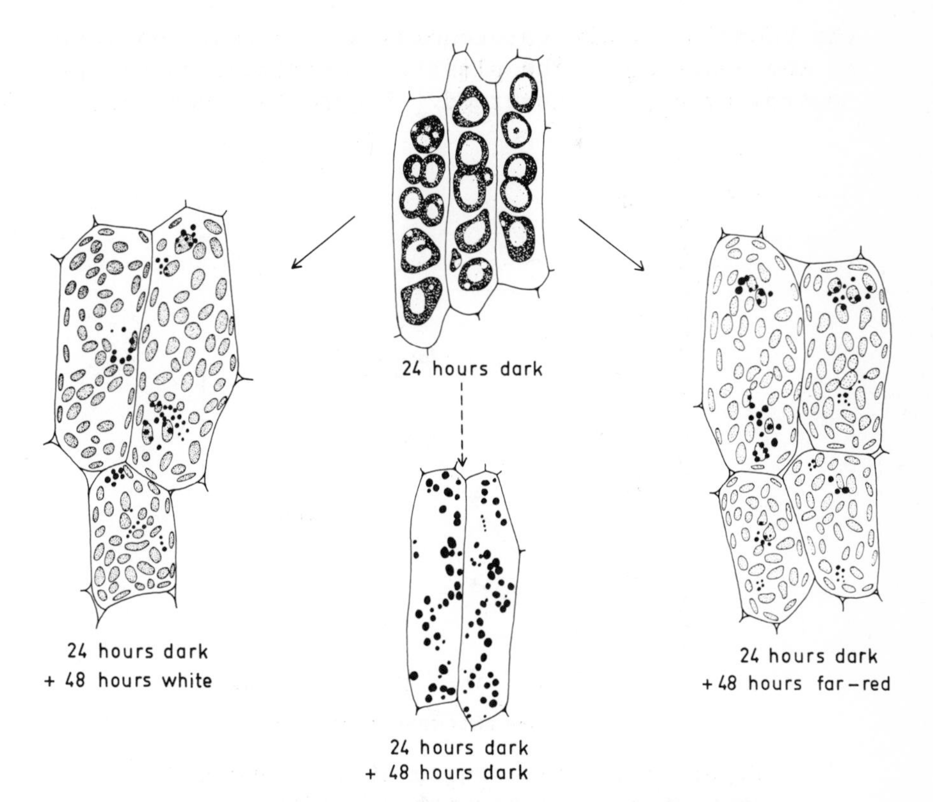

Figure 18. These drawings illustrate the control exerted by continuous standard far-red light (i.e., by P_{fr}) over degradation of storage protein (represented by dark, dense bodies 72 hr after sowing) and formation of plastids in mesophyll cells of mustard seedling cotyledons. The small etioplasts of the dark-grown seedlings were not drawn. Onset of light: 24 hr after sowing. (After Häcker, 1967.)

etioplasts. The structural substance is arranged as a prolamellar body from which emerge a considerable number of thylakoids.

Carotenoids are typical constituents of the plastids. These terpenoid molecules are restricted to the plastid compartment. The assumption can be made that the behavior of the carotenoids may be representative of the behavior of the plastid compartment as far as control by P_{fr} is concerned. Figure 19 shows that

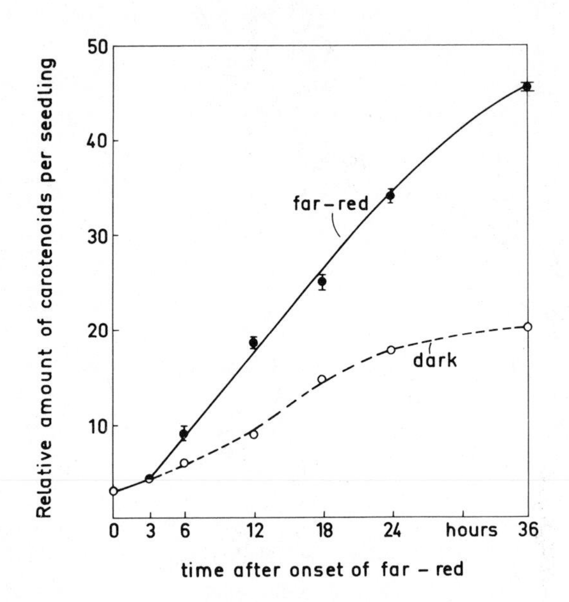

Figure 19. Time course of carotenoid accumulation in the mustard seedling in darkness and under the control of P_{fr}. Onset of standard far-red: 36 hr after sowing. After a lag-phase of about 3 hr, P_{fr} increases the rate of accumulation and prolongs the period of synthesis. (After Schnarrenberger and Mohr, 1967.)

carotenoid synthesis in the mustard seedling indeed responds to P_{fr} in much the same way as the structural unit "plastid." The initial lag-phase is 3 hr. Phytochrome-mediated plastid formation and carotenoid synthesis are strictly correlated. Neither phenomenon is correlated with chlorophyll synthesis.

Specific inhibitors of transcription and translation can easily be tested with respect to carotenoid synthesis. In spite of some effort in this direction, the answer to the question of how P_{fr} controls plastid formation and carotenoid synthesis remains vague. All data are consistent with the concept that P_{fr} increases the rate of activity of some genes involved in the growth and development of the plastid compartment. As a consequence the rate of carotenoid accumulation in the plastids is increased.

Since this state of affairs is not very satisfactory, we had better proceed to the next topic: phytochrome-mediated anthocyanin synthesis. This response proved to be a useful molecular model system for the analysis of phytochrome-mediated photomorphogenesis. A dark-grown, etiolated mustard seedling forms hardly any anthocyanin, whereas under continuous far-red, the seedling synthesizes a great amount of the red pigment. We find five different anthocyanins in the mustard seedling, all of which contain cyanidin (Fig. 20) as

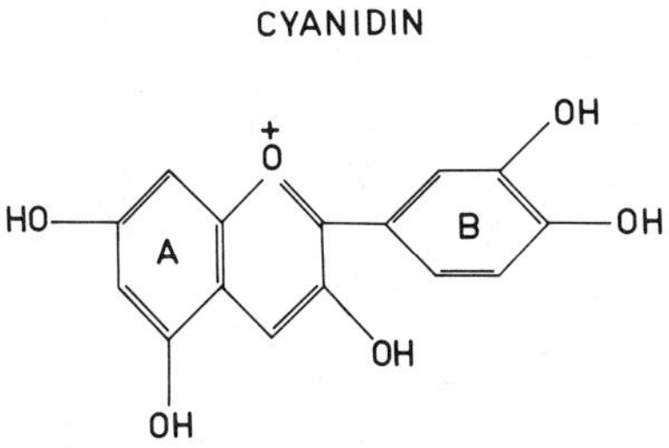

CYANIDIN

HO OH OH OH OH A B $\overset{+}{O}$

Figure 20. The formula of anthocyanidin cyanidin.

the colored aglycone. The problem is to describe in terms of molecular biology how P_{fr} mediates the synthesis of this molecule. It is usual in molecular work to know the course of a reaction before going on to investigate the conditions that control it. By and large in the case of cyanidin synthesis, this requirement is fulfilled. We know that the ring B and the heterocycle of the cyanidin molecule are derived from a phenylpropane skeleton, whereas the A ring originates from a head-tail condensation of C_2 units.

We recognize (Fig. 14) that for at least 20 hr after the initial lag-phase, the accumulation of anthocyanin is linear with time. The slope of the curve is a function of the quantum flux density of the far-red applied. One may conclude that, over this period of time, a virtually stationary concentration of P_{fr} will lead to a constant rate of anthocyanin synthesis.

Mainly on the basis of inhibitor experiments over the years, we have advanced the hypothesis that phytochrome-mediated anthocyanin synthesis is the result of a differential gene activation which is promoted in one way or another by P_{fr}. Figure 21 illustrates this hypothesis. P_{fr} is thought to cause in one way or another, through the activation of potentially active genes, the formation of one or several enzymes which are required for anthocyanin synthesis. This scheme has been verified by a great number of inhibitor experiments, but, as everyone knows, inhibitor experiments are often looked at which scepticism. Therefore, the direct measurement of phytochrome-mediated enzyme synthesis would be a desirable verification of the hypothesis that such synthesis is the basis of phytochrome-mediated anthocyanin synthesis. Pertinent experiments have been performed using L-phenylalanine ammonia-lyase, an enzyme which catalyzes the formation of trans-cinnamic acid from phenylalanine. This reaction is probably the first step of the biosynthetic sequence which leads from phenylalanine, which must be regarded as the main storage compound of phenolic nature in the mustard cotyledons, to the phenyl-propane moiety of the flavenoids, including the anthocyanidins (Fig. 22).

Phenylalanine ammonia-lyase activity, abbreviated PAL, can be induced by phytochrome in the mustard seedling. The results of the standard induction-reversion experiments are indicated in Table 2. Enzyme activity can be induced by "short-time irradiation" with red. The red effect can be reversed by immmediately following with far-red. The data show that most, if not all, of the total effect of 24 hr continuous red light can be attributed to P_{fr}. The effect of repeated, brief, far-red irradiations indicates that PAL synthesis is very sensitive toward P_{fr}. The photostationary state which is established by our standard far-red source maintains only 5% of the total phytochrome in the P_{fr} form.

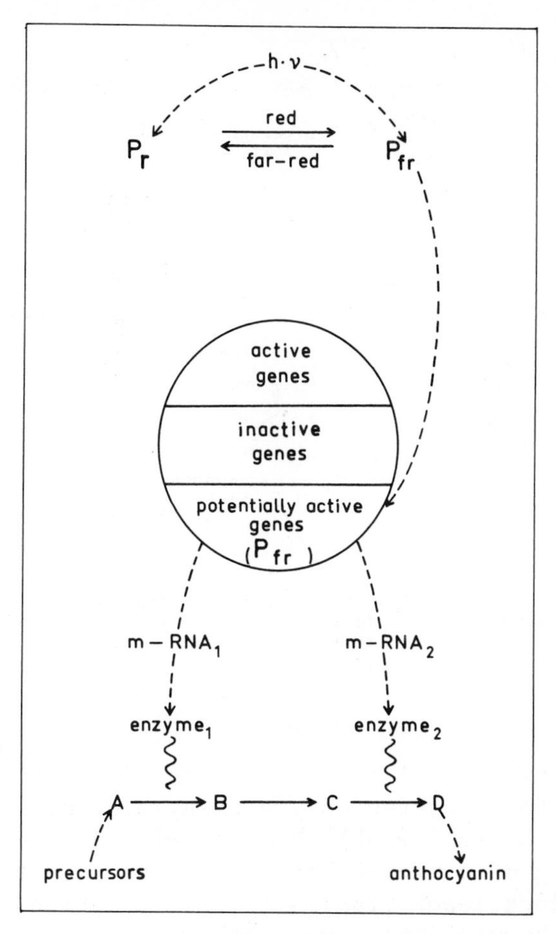

Figure 21. This scheme illustrates the hypothesis of differential gene activation, exerted in some way or another by P_{fr}, in the special case of phytochrome-mediated anthocyanin synthesis. The central dogma of molecular biology, DNA-RNA-protein, is taken for granted.

This sort of experiment can only give us the information that phytochrome is involved in the control of the enzyme level in the mustard seedling. Further analysis requires kinetics studies under steady-state conditions with respect to P_{fr}, that is, under continu-

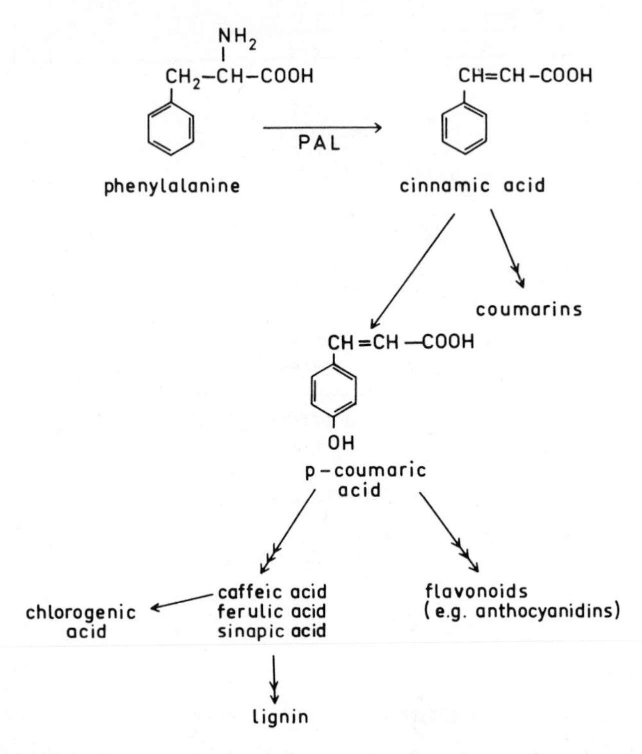

Figure 22. A sketch to emphasize the importance of the enzyme phenylalanine ammonia-lyase (PAL) which catalyzes the first step of a biosynthetic sequence. This sequence leads finally to the formation of anthocyanidins, among other compounds.

ous far-red. Figure 23 shows the kinetics of enzyme induction. The initial lag-phase after the onset of far-red is 1.5 hr, and the phytochrome-mediated increase of activity is sharp and rapid. The rate of increase is constant over many hours, with the activity remaining at a plateau for at least 6 hours and decreasing later on. We might notice this one point: If we accept the central dogma of molecular biology, namely the existence of the DNA-mRNA-enzyme relations, the sudden and linear increase of enzyme activity means that the mRNA involved has a rapid turnover, since a steady state concentration seems to be established

TABLE 2

Conventional induction-reversion experiments to demonstrate the involvement of phenylalanine ammonia-lyase (EC 4.3.1.5.) activity in the mustard seedling (*Sinapis alba* L.). This experiment can only give the information that phytochrome is involved. Further analysis requires kinetic studies under steady-state conditions with respect to P_{fr}. (After Weidner, Rissland, and Mohr, 1968.)

Program	enzyme activity $\left[\frac{\mu\text{mole trans-cinnamic acid}}{\text{min X seedling}}\right] \times 10^{-4}$
36 hr d*	0.11
36 hr d + 24 hr d	0.29
36 hr d + 24 hr red	0.68
36 hr d + 4 X (5 min red + 355 min d)	0.66
36 hr d + 4 X (5 min red + 5 min fr + 350 min d)	0.43
36 hr d + 4 X (5 min fr + 355 min d)	0.45

*d = dark; fr = far-red.

within a short period of time.

This conjecture looks promising, but we are left with the central problem of how the initial (or primary) lag-phase can be understood. What happens during the 1.5 hr which must pass after the first onset of light before enzyme synthesis can start? Can we shorten the lag-phase? The answer is yes, we can. If a seedling which has been irradiated for a period of 12 hr with far-red is transferred to darkness, enzyme synthesis stops virtually immediately (Fig. 24). Since there is no appreciable enzyme decay at this stage, the

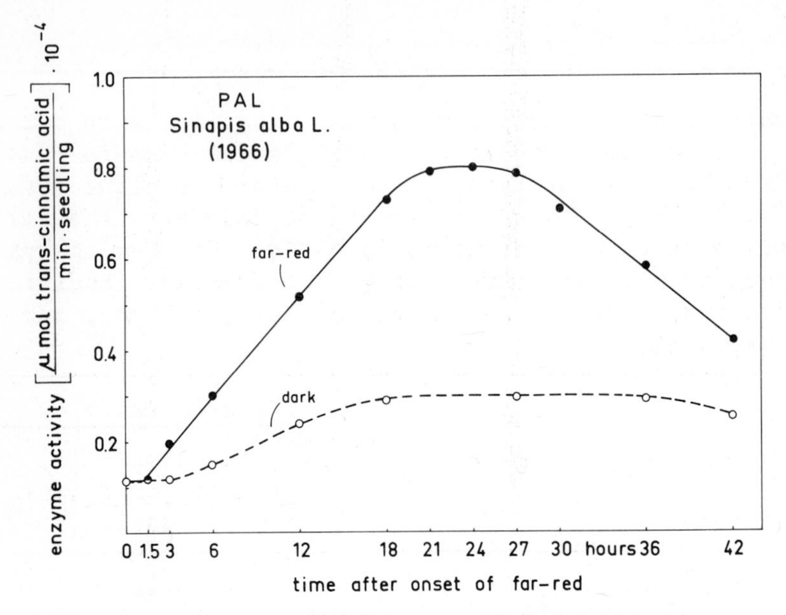

Figure 23. The induction of PAL activity in the mustard seedling by continuous standard far-red light. The initial (or primary) lag-phase of the response is 1.5 hr. (From experiments of Dr. M. Weidner, I. Rissland, and L. Lohmann.)

enzyme activity remains constant over a considerable period of time. If, for example, the seedling is kept in darkness for 6 hr and then re-irradiated with far-red, no lag-phase for the action resulting from the second irradiation (i.e., no secondary lag-phase) can be detected.

Since the action resulting from the second irradiation, as measured by increase of enzyme activity, can be completely inhibited by low doses of Puromycin and Cycloheximide, well-known inhibitors of protein synthesis, we conclude that the re-appearance of P_{fr} rapidly leads to *de novo* synthesis of enzyme protein.

In view of some confusion in the literature, I should like to emphasize that it is essential to continue the primary irradiation until steady-state conditions with respect to enzyme production are established. Only then can one be sure that all cells which are ca-

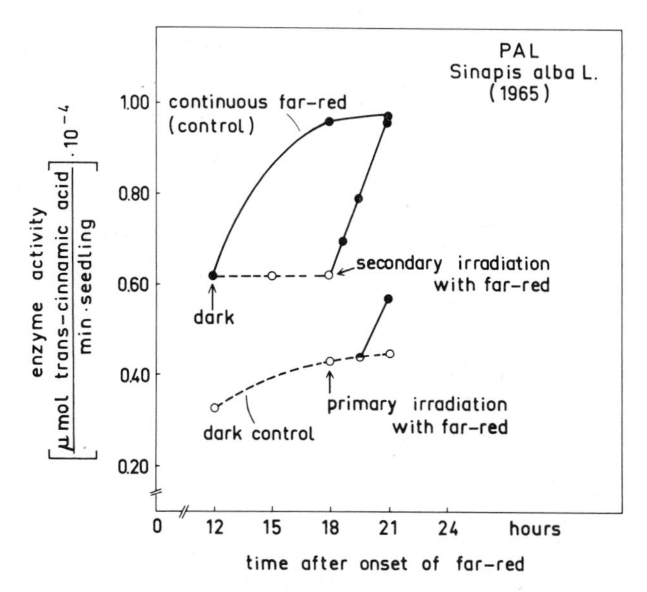

Figure 24. Initial and secondary lag-phases of far-red mediated increase of PAL activity in the mustard seedling. To determine the secondary lag-phase, the seedlings were irradiated for 12 hr with far-red, placed in darkness for 6 hr, and re-irradiated with far-red. (After Rissland and Mohr, 1967.)

pable of producing the enzyme under the given conditions are functioning. This is essential; otherwise one must expect a secondary lag-phase, because those cells which have not produced enzyme during the first irradiation will start enzyme production with a lag-phase after the onset of secondary irradiation.

Before we discuss the problem of how the initial lag-phase can be understood, we return briefly to investigate the problem of whether or not secondary lag-phases can be measured in anthocyanin synthesis (Fig. 25). The initial lag-phase after the onset of far-red is always approximately 3 hr. If, however, a seedling which was pre-irradiated with 12 hr of far-red is kept in darkness for an extended period and is then re-irradiated with far-red, no lag-phase for the action of the

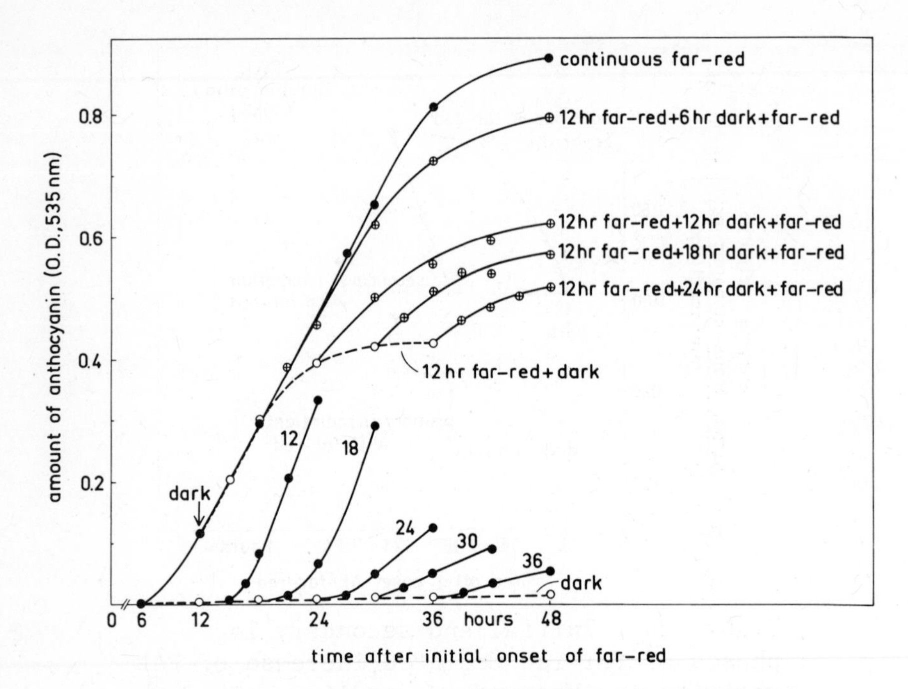

Figure 25. Initial lag-phases (lower part) and secondary lag-phases (upper part) of far-red mediated anthocyanin synthesis. To determine the secondary lag-phases, the mustard seedlings were irradiated for 12 hr with far-red, placed in darkness, and re-irradiated with far-red after dark intervals of different duration as indicated in the figure. The numbers (lower part of the figure) indicate the time of onset of far-red light. (After Lange, Bienger, and Mohr, 1967.)

second irradiation can be detected even when anthocyanin synthesis had already ceased during the preceding dark period. Since the action of the second irradiation as measured by anthocyanin synthesis can be completely inhibited by low doses of Puromycin and Cycloheximide, we conclude that the reappearance of P_{fr} leads to *de novo* synthesis of enzyme protein which is required for a resumption of anthocyanin synthesis.

These are the facts (Figs. 24 and 25): There is a

relatively long, initial lag-phase of several hours duration and no detectable secondary lag-phase.

A tentative interpretation of these facts can be summarized as follows: P_{fr} exerts two functions during the initial lag-phase. First, it makes in some way or another the potentially active genes accessible for the activating action of P_{fr}. Second, P_{fr} starts the transcription of potentially active genes. To maintain gene activity, the continuous presence of P_{fr} is required. On the other hand, if a gene has once been "opened" to the action of P_{fr}, it remains immediately accessible for the activation action of P_{fr} even in the case when P_{fr} has disappeared and gene activity has ceased for an extended period of time.

This sort of interpretation is supported by the finding that the sensitivity toward a standard dose of Actinomycin D, a potent inhibitor of transcription which combines with DNA, is altered under the influence of P_{fr}, and by the experimental evidence (Fig. 26) that the length of the initial lag-phase cannot be shortened by increasing the intensity of the standard far-red, in spite of the fact that the rate of anthocyanin synthesis after the lag-phase is a function of the quantum flux density applied. It is obvious that phytochrome action during the initial lag-phase and during the production phase do have quite different intensity dependencies. Our knowledge about the primary lag-phase is admittedly tenuous. Irrespective of the interpretation of the primary lag-phase, however, the conclusion seems to be fully justified that in the case of secondary irradiation, P_{fr} can rapidly cause differential enzyme synthesis.

Now I have to justify the term "differential." Some enzymes which have been studied in the mustard seedling in connection with photomorphogenesis are not under the control of phytochrome. One of these is isocitritase. This enzyme catalyzes the formation of glyoxylate and succinate from isocitrate. P_{fr} does not have any influence on the activity of isocitritase in the mustard seedling, although the enzyme activity does show a strong increase and a following decline during the period of experimentation (Fig. 27).

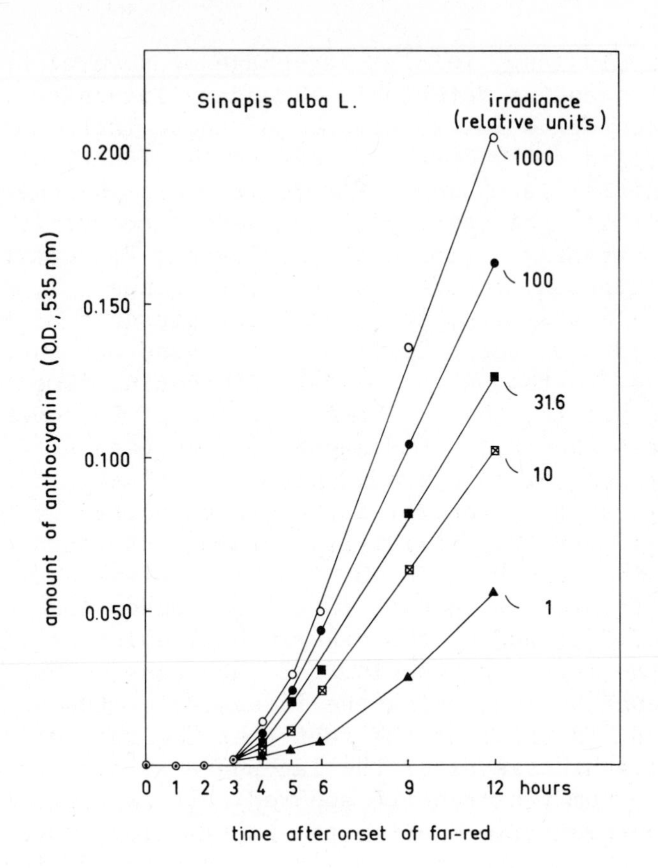

Figure 26. The kinetics of anthocyanin under continuous far-red light of different quantum flux densities. The photoequilibrium is always the same. Although the slopes differ, the lag-phase is not affected by irradiance. (From experiments of H. Lange.)

The behavior of isocitritase as compared with that of the phenylalanine ammonia-lyase supports the view that photomorphogenesis is not due to a general metabolic change in the cells; it rather seems to be due to very specific changes on the enzyme level.

You remember from a previous paragraph that we have used the term "negative" photoresponses for those phytochrome-mediated photoresponses which are character-

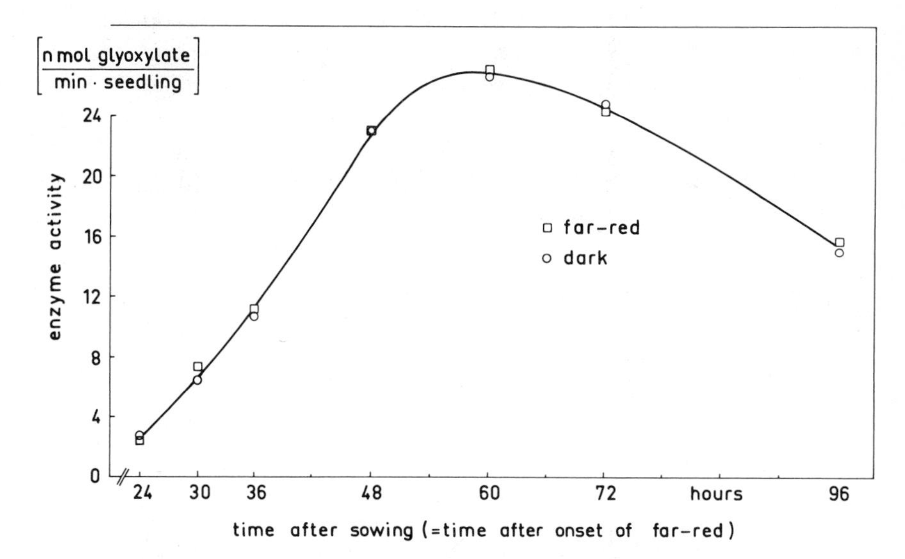

Figure 27. Time course of isocitritase (isocitrate-lyase) activity in the mustard seedling in darkness and under continuous standard far-red light. (After Karow and Mohr, 1966.)

ized by an inhibition of growth or metabolic processes. We have described the inhibition of hypocotyl lengthening as a typical response of this sort (Fig. 15). A characteristic of the negative photoresponses is that even the primary lag-phase is short.

The problem has been to study the causalities of "negative" photoresponses in molecular terms. So far, growth phenomena have proved too complex to be analyzed this way. Recent results obtained with the enzyme lipoxidase look more promising. Lipoxidase is widely distributed in the plant kingdom, but unfortunately, its physiological significance is still undertain. Nevertheless, it is a fine model system for the investigation of enzyme repression by P_{fr}. Unsaturated fatty acids containing a methylene-interrupted, multiple-unsaturated system in which the double bonds are all cis are oxidized in the presence of lipoxidase. Substrates conforming to this requirement include linoleic and linolenic acid. The overall reaction involves a peroxidation of the methylene-interrupted cis-cis acid to a conjugated cis-trans peroxide. Figure 28 shows that the increase of lipoxidase activity in the mustard seedling is arrested immediately after the onset of far-red. This fact can be interpreted as a repression of enzyme synthesis by P_{fr}.

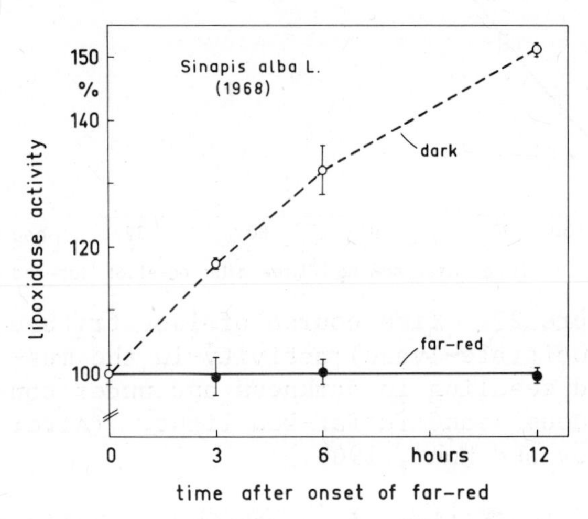

Figure 28. The increase of lipoxidase activity in the dark-grown mustard seedling is arrested by continuous standard far-red. The extrapolation indicates that the lag-phase is very short. (Standard errors are given in this particular case to justify the extrapolation.) (From experiments of H. Karow.)

The following speculative model (Fig. 29) summarizes the available molecular facts with respect to the action of P_{fr} on *morphogenesis* in a developing plant. It is now generally understood that all living cells of a particular plant (except perhaps the sieve tube elements) contain the total complement of genetical DNA that is characteristic for the individual. In

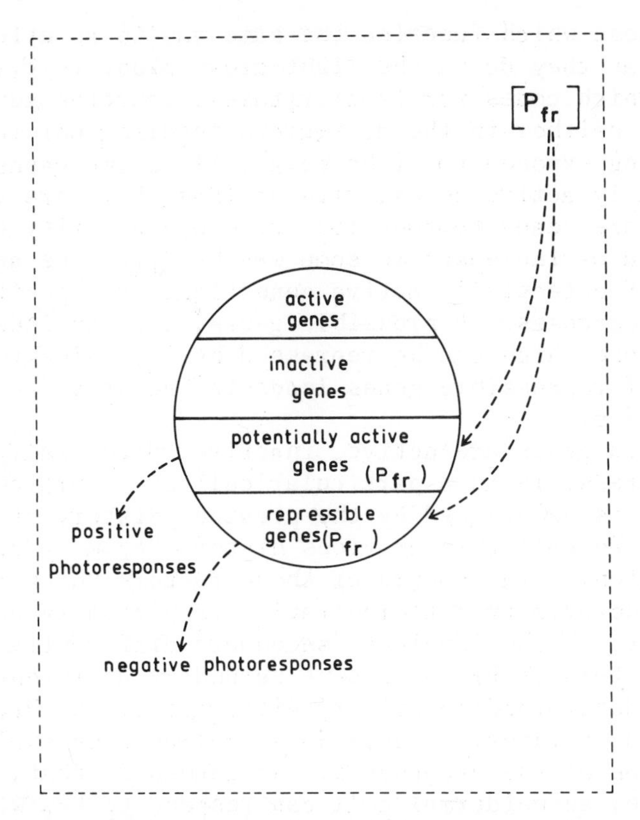

Figure 29. This scheme illustrates the hypothesis of differential gene activation and differential gene repression through P_{fr}. (The hypothesis is explained in some detail in the text.)

other words, all the genes are present in all cells, but only a fraction of them are active at a given time in a given cell. Genes are thus turned off and on. Such differential gene activity results in differences among cells that have the same set of genes. In connection with photomorphogenesis we have to refine this general scheme in the following way. The total genes of each particular cell of a dark-grown seedling which is able to respond to P_{fr} must be divided into at least four functional types: active, inactive, potentially active, and repressible genes. Active genes are those which function the same way in an etiolated plant as they do in the light-grown plant (e.g., the gene which codes for isocitritase), inactive genes are active neither in the dark-grown seedling nor in the seedling exposed to light (e.g., flowering genes). Potentially active genes, with an index P_{fr}, are those which are ready to function and whose activity can be started or increased in some way by P_{fr}. The activation of potentially active genes leads to "positive" photoresponses. Repressible genes, with an index P_{fr}, are those which can be repressed by P_{fr}. The repression of repressible genes leads to "negative" photoresponses.

Which genes are active, inactive, potentially active, or repressible in a particular cell at a particular moment is determined by the previous history of the cell. We call this previous history "primary differentiation." The nature of those factors which regulate the primary differentiation is virtually unknown. P_{fr} acts on the level of "secondary differentiation." The pattern of P_{fr}-dependent secondary differentiation is predetermined by primary differentiation; P_{fr} is only the trigger. Figure 30 serves as a general justification of the hypothesis. We recognize that, for example, an epidermal cell can respond to P_{fr} with the formation of a hair, a "positive" photoresponse, as well as with the reduction of the rate of lengthening, a "negative" photoresponse. A subepidermal cell can respond with the formation of anthocyanin as well as with the reduction of lengthening.

In our model the analogy with models developed in bacterial systems is obvious. One striking difference, however, between gene regulation in bacteria and in multicellular systems is that whereas in bacteria many of the effectors of gene action are directly related to, or identical with, the metabolites of the pathways they control, in multicellular systems the corresponding agents, the effector molecules, such as P_{fr} or hormones, are specialized molecules that are unrelated to the pathways which they control. However, in neither case is it fully known how an effector substance acts at the molecular level to modify the transcription

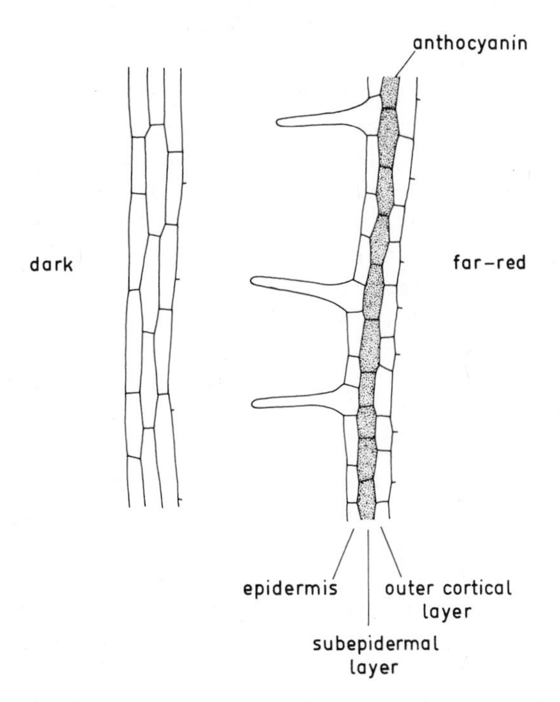

Figure 30. These drawings represent the three outer cellular layers of the hypocotyl of a mustard seedling. Left: from a dark-grown seedling. Right: from a seedling which was kept for some time under far-red light.

process on the DNA. Likewise the primary (or initial) action of P_{fr} is still a mystery. By the term "primary action," one designates the first reaction in which P_{fr} is involved after its formation. This reaction is completely obscure. However, in a number of instances (even in flowering) it has been shown that this reaction can occur in as little time as a minute. It is generally believed that the primary action of P_{fr} is the same in all cells. With the same confidence, however, one can maintain that P_{fr} acts differently in different cells or in different compartments of a particular cell.

I would like to reconsider the concept of the primary action of P_{fr}. As I understand it this concept envisions one primary act by P_{fr}, from which all other reactions follow in sequence. I would like to take the liberty of questioning this concept of the occurrence of one exclusive primary act as being too simplified. It is possible that phytochrome, although it is the same in all tissues, acts differently in different cells and in different compartments of the cell.

This problem leads me to the final point. Phytochrome is involved in a number of responses which are not related to morphogenesis and, very probably, not to gene activation. One example, previously described, is the phytochrome-mediated chloroplast movement in fully differentiated algal cells and the photonastic leaflet movement of the sensitive plant, *Mimosa pudica*. Both are rapid phenomena; under optimum conditions they can be fully expressed in 30 min. I will describe briefly the involvement of phytochrome in the photonastic movements of mature mimosa leaves. This response has been thoroughly investigated by the Beltsville group.

Movements about the pulvini of the leaves, pinnae, and pinnules of the sensitive plant in response to touch and variations in regimes of light and darkness have long attracted attention. We will follow the closing movements of pinnules on detached pinnae upon change from high-intensity fluorescent light to darkness (Fig. 31). This response depends upon the presence of phytochrome in the far-red absorbing P_{fr} form. In Fig. 31 pinnae are shown 30 min after transition from high-intensity fluorescent light to darkness. At the time of transition they were irradiated in succession for 2 min with red or far-red light to establish the corresponding photoequilibria. The pinnae remained open if exposure to far-red was last, and closed if red radiation was last. There is evidence from experiments with *Albizzia* that the tertiary pulvini, the sites of the response of the pinnules, act as the photoreceptors. The mimosa response has led to the conclusion that the P_{fr} action in the pulvini is on mem-

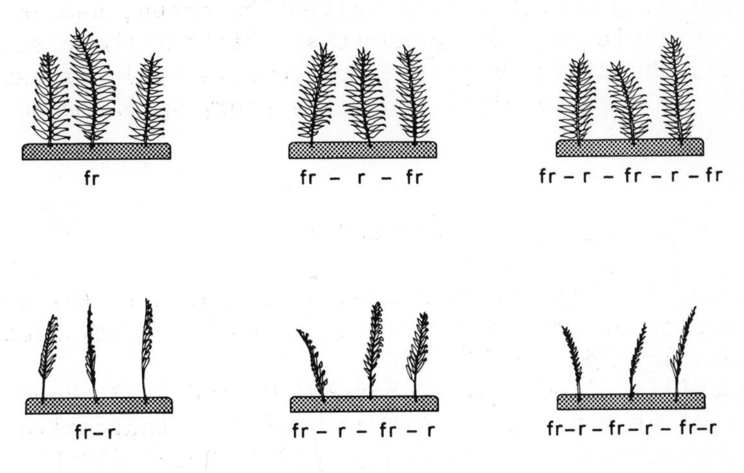

Figure 31. Pinnae of *Mimosa pudica* L., 30 min after transition from high-intensity fluorescent light to darkness. At the time of transition they had been irradiated in succession for 2 min with red or far-red light to establish phytochrome predominantly in the P_{fr} or P_r form. The pinnae remained open if exposure to far-red was last (top row) and closed if red radiation was last (bottom row). (After Fondeville, Borthwick, and Hendricks, 1966.)

brane permeability. It is sometimes speculated that P_{fr} action on membrane permeability is *the* "primary action" of P_{fr} in the previously defined sense. While it is indeed possible that this might be true for the pulvini cells, positive evidence that differential changes in membrane permeability are universally connected with phytochrome action is lacking. We must confess that at this time the *primary* action of P_{fr} is still a mystery.

I should like to close this article with a personal remark. I began to work with the mustard seedling about a decade ago as a postdoctoral fellow at Beltsville. It was a great privilege to be, at least temporarily, a member of the Beltsville group, headed by Dr. Borthwick and Dr. Hendricks. Without these scientists, the field of photomorphogenesis would probably still be phenomenology and not a story which makes sense.

REFERENCES

Etzold, H., Polarotropism and phototropism in the chloronemas of *Dryopteris filix-mas* (L.) Schott. *Planta*, *64*, 254-280 (1965).

Fondeville, J. C., H. A. Borthwick, and S. B. Hendricks, Leaflet movement of *Mimosa pudica* L. indicative of phytochrome action. *Planta*, *69*, 359-364 (1966).

Haecker, M., Phytochrome control of degradation of storage protein and formation of plastids in the cotyledons of the mustard seedling (*Sinapis alba* L.). *Planta*, *76*, 309-325 (1967).

Hartmann, K. M., Ein Wirkungsspektrum der Photomorphogenese unter Hochenergiebedingungen un seine Interpretation auf der Basis des Phytochroms (Hypokotylwachstumschemmung bei *Lactuca sative* L.). *Z. Naturforschg.*, *22b*, 1172-1175 (1967).

Jakobs, M., and H. Mohr, Kinetical studies on phytochrome-induced protein synthesis. *Planta*, *69*, 187-197 (1966).

Karow, H., and H. Mohr, Changes of activity of isocitritase (EC 4.1.3.1.) during photomorphogenesis in mustard seedlings. *Planta*, *72*, 170-186 (1967).

Lange, H., I. Bienger, and H. Mohr, New evidence in favor of the hypothesis of differential gene activation by phytochrome 730. *Planta*, *76*, 359-366 (1967).

Mohr, H., Differential gene activation as a mode of action of phytochrome 730. *Photochem. Photobiol.*, *5*,

469-483 (1966).

Rissland, I., and H. Mohr, Phytochrome-mediated enzyme formation (phenylalanine deaminase) as a rapid process. *Planta*, *77*, 239-249 (1967).

Schnarrenberger, C., and H. Mohr, Phytochrome-mediated synthesis of carotenoids in mustard seedlings (*Sinapis alba* L.). *Naturwiss.*, *54*, 837 (1967).

Weidner, M., I. Rissland, and H. Mohr, Photoinduction of phenylalanine ammonia-lyase in mustard seedlings: involvement of phytochrome. *Naturwiss.*, in press (1968).

Photochemistry and Biochemistry in Photosynthesis

M. Avron

Department of Biochemistry
Weizmann Institute of Science
Rehovot, Israel

Recent progress in our understanding of the process of photosynthesis stems to a large extent from the emergence of a cohesive picture of the mechanism of the photoinduced, electron transport reactions. The essence of this picture is the involvement of two separate photochemical reactions, operating in series, with chlorophyll serving as the major light-harvesting pigment of both photosystems.

The advantages inherent in such a scheme became clear soon after its first version was put forward by Hill and Bendall (1). During photosynthesis a strong reductant is photoproduced ("X") with a redox potential of around -0.6 v according to recent estimates (2, 3, 4). The electron-donating reaction is the slitting of water to evolve free oxygen with a redox potential of +0.8 v. This increase in the reducing power of the electron by 1.4 v can be accomplished by light of

680 mμ, each quantum of which has enough energy, assuming 100% efficiency, to increase the reducing power of an electron by 1.7 v. Thus, an energy utilization efficiency of 1.4/1.7 = 82% is required for the process to operate, an extremely stringent requirement at best, and, accepting some reasonable assumptions, an impossibility (5). Thus, the cooperation of two quanta in this photoinduced electron transfer is strongly suggested solely by energetic consideration.

However, the experimental evidence which has been accumulated and is in agreement with the general outline of the schematic presented in Fig. 1 is volumin-

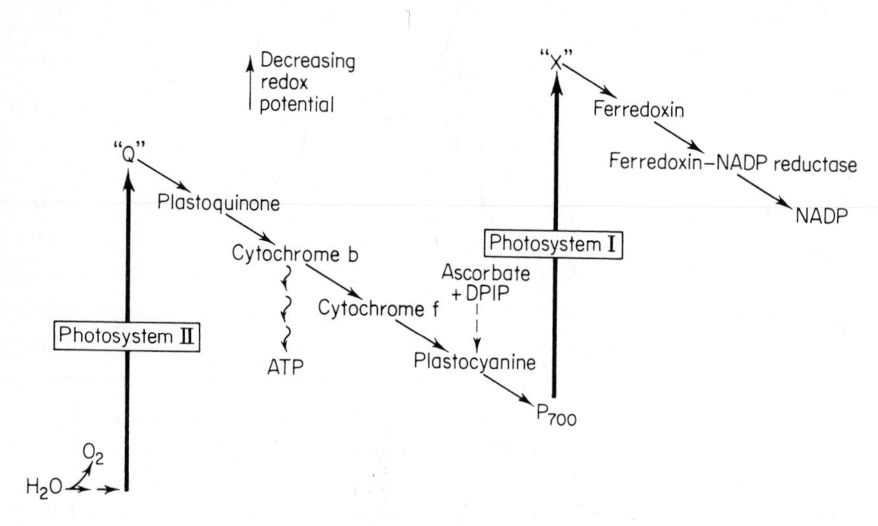

Figure 1. Schematic of photoinduced electron transport reactions in isolated chloroplasts.

ous, varied, and impressive (6). I will try to describe and illustrate some of the major lines of evidence. First, let me reiterate in words what the schematic of Fig. 1 is depicting. Starting on the left, the water serves as the electron donor in a photoreaction promoted by a chlorophyll-containing complex termed photosystem II. The electron acceptor is an unknown compound, called "Q", whose properties have been indirectly studied. Its redox potential is around 0.0 to + 0.18 v (7, 8). Thus, photosystem II promotes the transfer of an electron against a redox gradient of 0.6-0.8 v utilizing a quantum of light which has enough energy to move an electron against a gradient of 1.7 v. The reduced "Q" now transfers its electron, in the dark, through a series of membrane-bound components including plastoquinone, one or two cytochromes, a copper protein called plastocyanin, and, finally, to a component termed P_{700}. P_{700} is most probably a chlorophyll molecule which, due to not fully understood reasons, behaves in a manner different from the bulk chlorophyll (9). Like cytochrome f or plastocyanin, it is present in a ratio of about 1 per 500 molecules of bulk chlorophyll and undergoes oxidation-reduction accompanied by an easily observable absorption change. Its redox potential has been determined to be +0.4 v (9). Thus, an electron traversing the chain between the two photosystems loses the energy equivalent of about 0.2 to 0.4 v. This is enough, when properly coupled, to promote the formation of one to two ATP molecules from ADP and inorganic phosphate. The evidence presently available indeed suggests, as indicated in the figure, that at least one site of coupled phosphorylation is located in the electron transfer path between the two photosystems (10).

Reduced P_{700} now serves as the electron donor in a photoreaction promoted by another chlorophyll-containing complex termed photosystem I. The electron acceptor is an unknown compound called "X", whose redox potential has recently been determined to be about -0.6 v (2, 3, 4). Thus, photosystem I promotes the transfer of an electron against a redox gradient of 1.0 v. Via several electron carriers some of which have been isolated and characterized (6), reduced "X" reduces the final product of this electron transport schematic, NADP.

We have already mentioned the energetic considerations which lead to a requirement for cooperation of two photons of light in the photoreduction of NADP by water. However, the first experimental observation which indicated that the process was not a simple photoreaction sensitized by chlorophyll was made by Emerson and Lewis in 1943 (11). They observed that the quantum yield of photosynthesis was reasonably constant between 500 and 680 mμ, but dropped severely beyond 680 mμ. Since chlorophyll a is the only major light absorber in the latter region it was clear that light absorption by chlorophyll alone was insufficient for efficient photosynthesis to proceed. This phenomenon, termed the "red-drop," can be interpreted, in view of present knowledge, as due to the location of a specialized type of chlorophyll, "long-wavelength absorbing chlorophyll," exclusively in photosystem I. Since equal excitation of both photosystems is required for photosynthesis to proceed (See Fig. 1), it is clear why the efficiency considerably decreases when one photosystem absorbs most of the light.

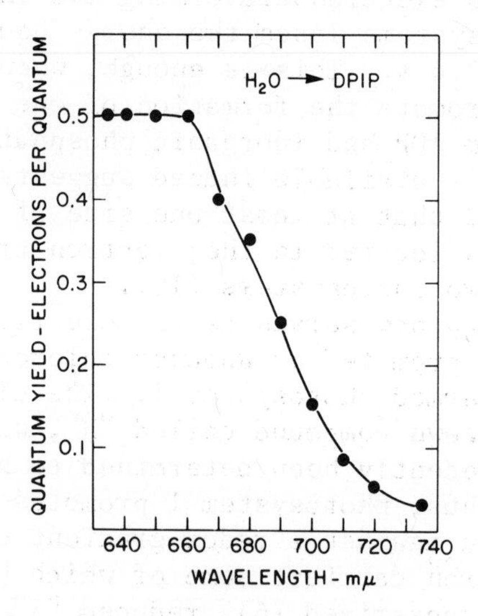

Figure 2. The red-drop phenomenon as recently observed in the photoreduction of dichlorophenol-indophenol by isolated chloroplasts. (After Sauer and Kelly, 1965.)

This interpretation of the red-drop phenomenon was

strongly supported by the further observation of Emerson and his collaborators (12) that the low efficiency of photosynthesis beyond 680 mμ could be considerably improved if, simultaneously, illumination with a shorter wavelength was provided. Thus, the low efficiency of the far-red (beyond 680 mμ) light, due to its being absorbed mostly by photosystem I, could be improved when one supplied the system simultaneously with shorter wavelength light. This light was mostly absorbed by photosystem II. The extent of this effect, termed the "enhancement" phenomenon, can most easily be determined (13) by utilizing the relation

$$\text{Enhancement} = \frac{\text{Rate (red + far-red)} - \text{Rate (red)}}{\text{Rate (far-red)}}.$$

Values exceeding 1 indicate the presence of enhancement and this the cooperation of both photosystems in the process studied. Recently it was observed that under conditions where the photoreduction of NADP by water showed considerable enhancement (14), that of ferricyanide by water did not (15) (Table 1). Therefore, it was concluded that in the latter photoreduction, only photosystem II participates; and ferricyanide must accept most of its electron in between the two photosystems.

TABLE 1

Enhancement in NADP and ferricyanide photoreduction by chloroplasts (After Avron and Ben-Hayyim, 1968)

Electron acceptor	Enhancement
NADP	2.5 ± 0.1
Ferricyanide	0.9 ± 0.1

A further, most crucial, independent support of the concept of two photochemical reactions operating in series came from the demonstration by Hoch and Martin

(16) that if one studies a reaction catalyzed by only photosystem I, the theoretical quantum yield of 1 electron per quantum absorbed is approached when excitation is limited to the "long-wavelength absorbing chlorophyll." They studied the photoinduced transfer of electrons from ascorbate to NADP (See Fig. 1). It was known from previous experiments (17) that ascorbate in the presence of a catalytic amount of dichlorophenol-indophenol feeds electrons into the chain in a position closer to NADP than does water, since only the latter process was inhibited by a variety of reactions, notably DCMU [(3,(3,4-dichlorophenyl)-1,1-dimethylurea)]. It was therefore considered plausible that it may feed its electrons in-between the two photosystems, and the transfer of electrons from ascorbate to NADP may require excitation of only photosystem I. As can be seen in Fig. 3, this indeed turned out to be the case.

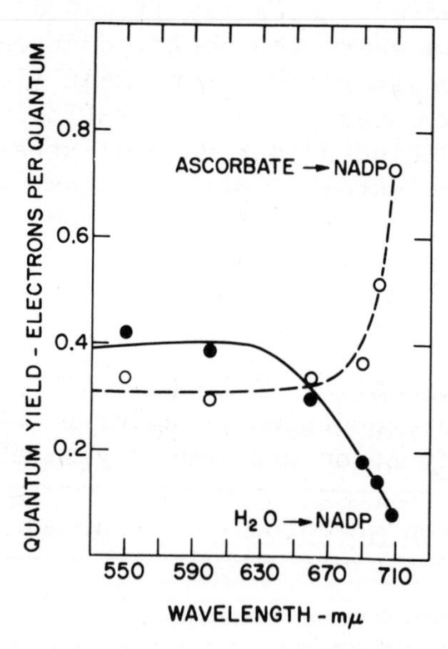

Figure 3. The quantum yield of the photoreduction of NADP from water or from ascorbate and dichlorophenol-indophenol in isolated chloroplasts. (After Sauer and Biggins, 1965.)

Whereas the electron transfer from water to NADP shows, as expected, the red-drop phenomenon, the transfer from ascorbate to NADP shows a "red-rise." Thus, since most of the "long-wavelength absorbing chlorophyll" is located in photosystem I, any reaction which requires the participation of photosystem II will exhibit a low efficiency in the long-wavelength region, while any reaction requiring only photosystem I will show improved efficiency in the region.

One difficulty remains if one accepts the presented picture. If indeed, one requires equal excitation of both photosystems for maximal efficiency, as is required by the schematic of Fig. 1, and on the other hand, the data just discussed indicates that photosystem I absorbs most of the light in the long-wavelength region while system II absorbs the larger proportion of light in the short-wavelength region (500-680 mμ), as indicated by the enhancement experiments, how is it possible to obtain a constant and theoretical quantum yield of one oxygen molecule evolved per eight quanta absorbed (one electron per two quanta) in the short-wavelength region, as observed by many workers in the field (18). Two types of theories have been suggested to explain this contradiction. The first, termed the "separate package hypothesis," assumes that the light-harvesting chlorophyll is almost equally distributed between the two systems, with photosystem II having only a slight edge. It is supported by the observation that to obtain maximal enhancement, high ratios of short-wave to long-wave light must be provided (13). Thus the concentration of the long-wavelength absorbing chlorophyll in photosystem I is considerably higher than the concentration of the light-harvesting, short-wavelength absorbing chlorophyll in photosystem II. The alternative or "spillover" hypothesis assumes that considerably more of the light-harvesting chlorophyll molecules are located in photosystem II than in photosystem I, but that a spillover mechanism exists whereby excess excitation energy absorbed in photosystem II can be funnelled into the photosystem I reaction center (but not vice versa). This hypothesis is supported by a mathematical model (19) of two photoreactions operating in series, which when applied to available data indicates the existence of considerable spillover. It is also strongly favored by the recent observations (15) that with some photoreactions involving photosystem II, such as the photoreduction of ferricyanide by water, a quantum efficiency of 1 has been obtained with short-wavelength light; and the same quantum yield with the same light was observed with another involving only photosystem I, the photoreduction of a viologen by ascorbate. Therefore, short-wavelength light could be directed essentially fully toward photosystem II, or toward photosystem I, depending upon the photoreaction being studied. This was interpreted as indicating that by far most of the light-harvesting chlorophyll molecules were located in photosystem II, but that under proper circumstances, essentially complete spillover of light energy absorbed in these chlorophyll molecules into the photosystem I reaction center was possible.

One of the critical experiments which contributed to the establishment of the series formulation was the demonstration by Duysens and his collaborators (20) that cytochromes bound within chlorophyll-containing membranes of photosynthetic cells were oxidized when illumination was limited to far-red light (excitation of mostly photosystem I), and rereduced when the illumination was with shorter wavelength light (excitation of mostly photosystem II). Thus, the cytochromes must be located between the two photosystems (Fig. 1). Figure 4 illustrates a similar experiment conducted with isolated chloroplasts where illumination of the chloroplasts with 730 mμ light induces complete oxidation of the internally bound cytochrome f, and a further illumination with 640 mμ light rereduces the cytochrome (21). Similar responses were also observed in the oxidation and reduction of P_{700} (9).

Physical separation of chloroplasts into two subfractions has been recently achieved (22). One of these fractions could promote only photosystem I sensitized reactions, such as the photoinduced electron transfer from ascorbate to NADP, and was enriched in P_{700} and cytochrome f. The other could promote only

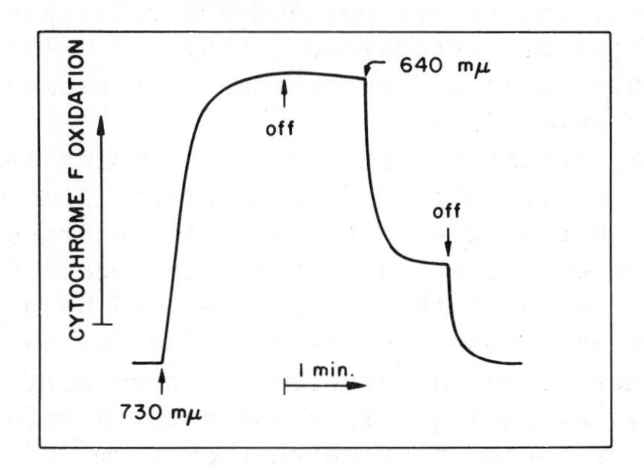

Figure 4. The oxidation of cytochrome f by far-red light and its reduction by red light in isolated chloroplasts. (After Avron and Chance, 1966.)

photosystem II sensitized reactions, such as the transfer of electrons from water to ferricyanide, and was enriched in a b-type cytochrome (22).

Finally, a long series of mutants lacking single components of the electron-transfer chain have been isolated (23, 24). Among them, mutants lacking cytochrome f, plastocyanin, or P_{700} have been clearly identified. The properties of these mutants are in agreement with those predicted from a linear series formulation, such as that presented in Fig. 1.

An established basic concept that had to be incorporated into present thinking, was that of the photosynthetic unit. This concept was originally suggested to explain the elegant, flash-yield experiments of Emerson and Arnold (25), in which it was shown that the maximum obtainable yield of a short, intense flash of light was one molecule of oxygen evolved per approximately 2500 chlorophyll molecules. Alternatively, one electron was transported per 600 chlorophyll molecules. Thus, when all the chlorophyll was excited rapidly, no more than one oxidation-reduction step could take place per 600 chlorophylls. This number turned out to be very popular since several electron transport components of chloroplasts were found to be present at about the same proportion, of one per 400-600 chlorophyll molecules. These are cytochrome f (26), P_{700} (9), and plastocyanin (27); and cytochrome b is present at about twice that level (28).

However, recent evidence seems to point back to the original, larger unit of 2500 chlorophyll molecules as the basic unit of the photosynthetic apparatus. Recent experiments by Joliot and his collaborators (29) in which they measured the oxygen produced by a single, short, intense flash disclosed that under such conditions no more than one electron is transported per 2800 chlorophyll molecules. They suggest, therefore, the existence of a component which they term "E", present at about the latter proportion. This suggestion is supported by several other independent observations. Izawa and Good carefully determined the minimal number of potent inhibitor molecules which must bind to chloroplasts to induce complete inhibition of oxygen evolution. Using several inhibitors, they arrived at a number of one molecule of inhibitor per 2500 chlorophyll molecules (30).

Finally, a rather potent tool for an indirect study of the electron transport path around photosystem II is provided by the changes in fluorescence yield of chloroplasts. It has long been known (31) that the fluorescence yield of photosynthetic cells increases by two to four-fold when the electron transport processes cease to operate (31). This is true when the electron transport stops because of a lack of substrate (CO_2), addition of an inhibitor, or a lack of an electron transport component of the cells. It was observed by several independent workers (32, 33, 34, 35) that in the absence of an electron acceptor system, the number of quanta of short wave (green) light that must be absorbed by the system to reach the maximum fluorescence yield is approximately 1/50 the chlorophyll concentration. Thus, electron acceptors must be naturally present to at least that extent. Further, a kinetic analysis of the time course of the rise in fluorescence yield coupled with low temperature studies indicate that this pool of electron acceptors is composed of two components of equal concentration. Since they have been shown to be oxidized by far-red illumination, and rereduced by short-wave illumination, they are generally considered to occupy the position of "Q" and plastoquinone in Fig. 1. When a potent inhibitor, such as DCMU, is added to block electron transport, a much smaller number of quanta must be absorbed to reach the maximum fluorescence yield. The numbers quoted by various investigators vary, but all indicate a component present at very low concentration, possibly as low as 1 per thousand chlorophylls. This again points to the presence of a large photosynthetic unit.

The schematic of Fig. 5 is an attempt to reconstruct

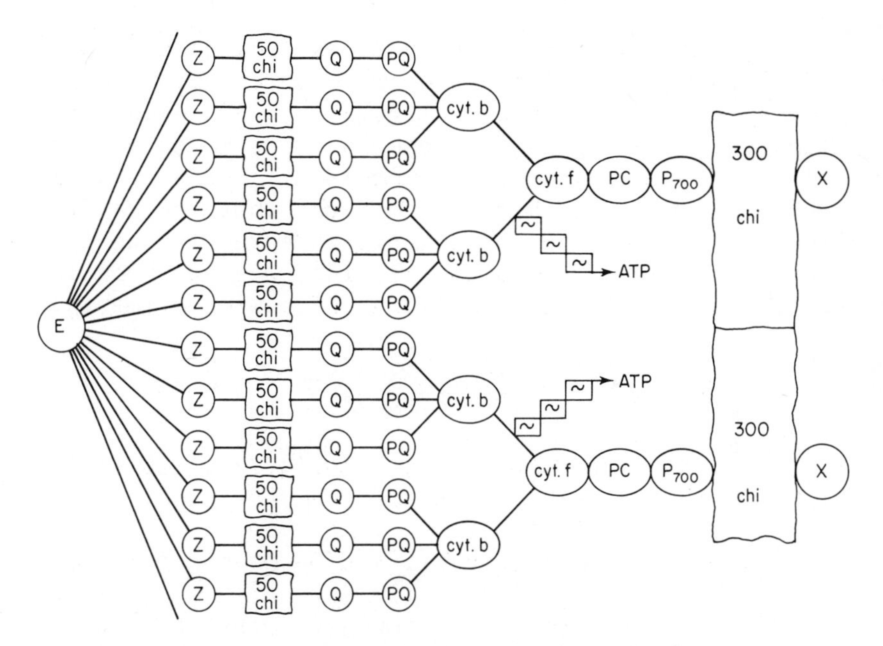

Figure 5. A hypothetical model of the photosynthetic unit in isolated chloroplasts. Only half of a unit is illustrated.

a model of a photosynthetic unit. It is composed of four identical 600 chlorophyll-containing subunits (only two are illustrated). Each of these subunits contains one cytochrome f, one plastoquinone, one P_{700}, two cytochrome b molecules, and six plastoquinones, "Q" and "Z" molecules. All are linked together into one "bottleneck" on the water side of photosystem II at "E", which is present at a concentration of about one per 2400 chlorophyll molecules and constitutes the site responsible for the measurements which indicate the large photosynthetic unit. The equal distribution of chlorophyll between the two photosystems, as indicated in the figure, is, of course, contrary to the recent data discussed earlier, which indicates that the majority of the light-harvesting chlorophyll is located in photosystem II. However, such redistribution will not affect the general features of the model as presented.

In conclusion, many observations from photochemical and biochemical approaches seem to agree with the general schemata presented in Figs. 1 and 5. Although agreement with regard to these is by no means universal and several alternative models are being considered even now (36, 37), there is no doubt that the unification brought into photosynthetic research by the introduction of these models has provided a great boon in serving as a guide to the design of new experiments.

REFERENCES

1. Hill, R., and F. Bendall, *Nature*, *186*, 136 (1960).
2. Zweig, G., and M. Avron, *Biochem. Biophys. Res. Comm.*, *19*, 397 (1965).
3. Kok, B., H. J. Rurainski, and O. Owens, *Biochim. Biophys. Acta*, *109*, 347 (1965).
4. Black, C. C., *Biochim. Biophys. Acta*, *120*, 332 (1966).
5. Duysens, L. N. M., *Plant Physiol.*, *37*, 407 (1962).
6. Avron, M., in *Current Topics in Bioenergetics*, D. R. Sanadi, ed. New York: Academic Press, Vol. II, 1 (1967).
7. Kok, B., S. Malkin, O. Owens, and B. Forbush, *Brookhaven Symposium on Biology*, *19*, 446 (1966).
8. Butler, W. L., *Proc. Int. Congr. of Photosynthesis*, Freudenstadt, Germany, in press (1968).

9. Kok, B., in *Plant Biochemistry*, J. Bonner and J. E. Varner, eds. New York: Academic Press, p. 903 (1965).
10. Avron, M., and J. Neumann, *Ann. Rev. Plant Physiol.*, *19*, 137 (1968).
11. Emerson, R., and C. M. Lewis, *Amer. J. Bot.*, *30*, 165 (1943).
12. Emerson, R., R. V. Chalmers, and C. Cederstrand, *Proc. Natl. Acad. Sci.*, *43*, 133 (1957).
13. Myers, J., in *Photosynthetic Mechanisms of Green Plants*, National Research Council Publication 1145, p. 301 (1963).
14. Govinjee, R., Govinjee, and G. Hoch, *Plant Physiol.*, *39*, 10 (1964).
15. Avron, M., and G. Ben-Hayyim, *Proc. Int. Congr. of Photosynthesis*, Freudenstadt, in press (1968).
16. Hoch, G., and I. Martin, *Arch. Biochem. Biophys.*, *102*, 430 (1963).
17. Vernon, L. P., and W. S. Zaugg, *J. Biol. Chem.*, *235*, 2728 (1960).
18. Kok, B., in *Encyclopedia of Plant Physiology*, A. Pirson, ed. Berlin: Springer Verlag, p. 566 (1960).
19. Malkin, S., *Biophys. J.*, *7*, 629 (1967).
20. Duysens, L. N. M., and J. Amesz, *Biochim. Biophys. Acta*, *64*, 243 (1962).
21. Avron, M., and B. Chance, in *Currents in Photosynthesis*, J. B. Thomas and J. C. Goedheer, eds. Rotterdam: Ad. Donker, p. 455 (1966).
22. Boardman, N. K., and J. M. Anderson, *Biochim. Biophys. Acta*, *143*, 187 (1967).
23. Givan, A. L., and R. P. Levine, *Plant Physiol.*, *42*, 1264 (1967).
24. Bishop, N. I., *Ann. Rev. Plant Physiol.*, *17*, 185 (1966).
25. Emerson, R., and W. J. Arnold, *J. Gen. Physiol.*, *16*, 191 (1932).
26. Davenport, H. E., and R. Hill, *Proc. Roy. Soc.*, *B19*, 327 (1952).
27. Katoh, S., I. Suga, I. Shiratori, and A. Takamiya, *Arch. Biochem. Biophys.*, *94*, 136 (1961).
28. James, W. O., and R. M. Leech, *Proc. Roy. Soc.*, *B160*, 13 (1964).

29. De Kouchkovsky, Y., and P. Joliot, *Photochemistry and Photobiology*, *6*, 567 (1967).
30. Izawa, S., and N. E. Good, *Biochim. Biophys. Acta*, *102*, 20 (1965).
31. Butler, W. L., in *Current Topics in Bioenergetics*, D. R. Sanadi, ed. New York: Academic Press, Vol. I, p. 49 (1966).
32. Duysens, L. N. M., *Progress in Biophys.*, *14*, 1 (1964).
33. Murata, N., M. Nishimura, and A. Takamiya, *Biochim. Biophys. Acta*, *120*, 23 (1966).
34. Malkin, S., and B. Kok, *Biochim. Biophys. Acta*, *126*, 413 (1966).
35. Joliot, P., *Biochim. Biophys. Acta*, *102*, 135 (1965).
36. Arnon, D. I., *Experientia*, *22*, 1 (1966).
37. Frank, J., and J. L. Rosenberg, *J. Theoretical Biol.*, *7*, 276 (1964).
38. Sauer, K., and R. B. Kelly, *Biochem.*, *4*, 2791 (1965).
39. Sauer, K., and J. Biggins, *Biochim. Biophys. Acta*, *102*, 55 (1965).